Supercritical Fluid Extraction and Chromatography

ACS SYMPOSIUM SERIES **366**

Supercritical Fluid Extraction and Chromatography
Techniques and Applications

Bonnie A. Charpentier, EDITOR
The Procter & Gamble Company

Michael R. Sevenants, EDITOR
The Procter & Gamble Company

Developed from a symposium sponsored
by the Division of Agricultural and Food Chemistry
at the 193rd Meeting
of the American Chemical Society,
Denver, Colorado,
April 5-10, 1987

American Chemical Society, Washington, DC 1988

Library of Congress Cataloging-in-Publication Data

Supercritical fluid extraction and chromatography: techniques and applications
   Bonnie A. Charpentier, editor, Michael R. Sevenants, editor.

   p.   cm.—(ACS symposium series, ISSN 0097–6156; 366)

   Developed from a symposium sponsored by the Division of Agricultural and Food Chemistry at the 193rd Meeting of the American Chemical Society, Denver, Colorado, April 5–10, 1987.

   Includes bibliographies and indexes.

   ISBN 0–8412–1469–7
   1. Supercritical fluid chromatography—Congresses.
   2. Supercritical fluid extraction—Congresses.

   I. Charpentier, Bonnie A., 1952–   . II. Sevenants, Michael R., 1938–   . III. American Chemical Society. Division of Agricultural and Food Chemistry. IV. American Chemical Society. Meeting (193rd: 1987: Denver, Colo.). V. Series

QD79.C45S87   1988
543'.0894—dc19   88–3466
                                               CIP

Copyright © 1988

American Chemical Society

All Rights Reserved. The appearance of the code at the bottom of the first page of each chapter in this volume indicates the copyright owner's consent that reprographic copies of the chapter may be made for personal or internal use or for the personal or internal use of specific clients. This consent is given on the condition, however, that the copier pay the stated per-copy fee through the Copyright Clearance Center, Inc., 27 Congress Street, Salem, MA 01970, for copying beyond that permitted by Sections 107 or 108 of the U.S. Copyright Law. This consent does not extend to copying or transmission by any means—graphic or electronic—for any other purpose, such as for general distribution, for advertising or promotional purposes, for creating a new collective work, for resale, or for information storage and retrieval systems. The copying fee for each chapter is indicated in the code at the bottom of the first page of the chapter.

The citation of trade names and/or names of manufacturers in this publication is not to be construed as an endorsement or as approval by ACS of the commercial products or services referenced herein; nor should the mere reference herein to any drawing, specification, chemical process, or other data be regarded as a license or as a conveyance of any right or permission to the holder, reader, or any other person or corporation, to manufacture, reproduce, use, or sell any patented invention or copyrighted work that may in any way be related thereto. Registered names, trademarks, etc., used in this publication, even without specific indication thereof, are not to be considered unprotected by law.

PRINTED IN THE UNITED STATES OF AMERICA

ACS Symposium Series

M. Joan Comstock, *Series Editor*

1988 ACS Books Advisory Board

Harvey W. Blanch
University of California—Berkeley

Malcolm H. Chisholm
Indiana University

Alan Elzerman
Clemson University

John W. Finley
Nabisco Brands, Inc.

Natalie Foster
Lehigh University

Marye Anne Fox
The University of Texas—Austin

Roland F. Hirsch
U.S. Department of Energy

G. Wayne Ivie
USDA, Agricultural Research Service

Michael R. Ladisch
Purdue University

Vincent D. McGinniss
Battelle Columbus Laboratories

Daniel M. Quinn
University of Iowa

E. Reichmanis
AT&T Bell Laboratories

C. M. Roland
U.S. Naval Research Laboratory

W. D. Shults
Oak Ridge National Laboratory

Geoffrey K. Smith
Rohm & Haas Co.

Douglas B. Walters
National Institute of
Environmental Health

Wendy A. Warr
Imperial Chemical Industries

Foreword

The ACS SYMPOSIUM SERIES was founded in 1974 to provide a medium for publishing symposia quickly in book form. The format of the Series parallels that of the continuing ADVANCES IN CHEMISTRY SERIES except that, in order to save time, the papers are not typeset but are reproduced as they are submitted by the authors in camera-ready form. Papers are reviewed under the supervision of the Editors with the assistance of the Series Advisory Board and are selected to maintain the integrity of the symposia; however, verbatim reproductions of previously published papers are not accepted. Both reviews and reports of research are acceptable, because symposia may embrace both types of presentation.

Contents

Preface ix

1. **Physical Chemistry of Supercritical Fluids: A Tutorial** 1
   C. T. Lira

2. **Processing with Supercritical Fluids: Overview and Applications** 26
   Val J. Krukonis

3. **Analytical Supercritical Fluid Extraction Methodologies** 44
   Bob W. Wright, John L. Fulton, Andrew J. Kopriva, and Richard D. Smith

4. **Supercritical Fluid–Adsorbate–Adsorbent Systems: Characterization and Utilization in Vegetable Oil Extraction Studies** 63
   Jerry W. King, Robert L. Eissler, and John P. Friedrich

5. **Concentration of Omega-3 Fatty Acids from Fish Oil Using Supercritical Carbon Dioxide** 89
   S. S. H. Rizvi, R. R. Chao, and Y. J. Liaw

6. **Supercritical Carbon Dioxide Extraction of Terpenes from Orange Essential Oil** 109
   F. Temelli, R. J. Braddock, C. S. Chen, and S. Nagy

7. **Steps To Developing a Commercial Supercritical Carbon Dioxide Processing Plant** 127
   R. T. Marentis

8. **Capillary Supercritical Fluid Chromatography with Applications in the Food Industry** 144
   T. L. Chester, L. J. Burkes, T. E. Delaney,
   D. P. Innis, G. D. Owens, and J. D. Pinkston

9. **Retention Processes in Supercritical Fluid Chromatography** 161
   Clement R. Yonker and Richard D. Smith

10. **Capillary Supercritical Fluid Chromatography: Use for the Analysis of Food Components and Contaminants** 179
    B. E. Richter, M. R. Andersen, D. E. Knowles,
    E. R. Campbell, N. L. Porter, L. Nixon,
    and D. W. Later

11. **Capillary Supercritical Fluid Chromatography–Mass Spectrometry: Practical Considerations and Applications** 191
    G. D. Owens, L. J. Burkes, J. D. Pinkston, T. Keough,
    J. R. Simms, and M. P. Lacey

12. **Supercritical Fluid Chromatography–Mass Spectrometry of Carotenoid Pigments** 208
    Nelson M. Frew, Carl G. Johnson, and
    Richard H. Bromund

13. **Supercritical Fluid Chromatography with Fourier Transform Infrared Detection** 229
    Richard C. Wieboldt and James A. Smith

INDEXES

Author Index 245

Affiliation Index 245

Subject Index 245

Preface

ONE OF THE MOST PERVASIVE PROBLEMS in chemistry is the separation of complex materials into component parts, either to characterize a mixture or natural product or to remove a particular component from a matrix. Supercritical fluids—in particular, supercritical carbon dioxide—are currently undergoing exciting developments in meeting both types of challenges while being used in extractions and in chromatography.

Supercritical fluid extraction (SFE) and supercritical fluid chromatography (SFC) are often considered disparate fields, with SFE being largely the purview of engineers and SFC that of analytical chemists. Both techniques are used in the analysis and processing of foods. Therefore, a symposium was held (from which this book was developed), in which learnings from both fields were combined to provide a useful summary of current trends and applications in the use of supercritical fluids in food-related research. Contributors (including leading researchers) from academia, government, and industry were chosen carefully to encompass different perspectives and areas of expertise in these fields.

Combining the contributions of these authors in one text provides a coherent mix of theory, history, state-of-the-art technology, and practical and commercial applications. We thank the authors for their participation and contributions. We also thank the ACS Books Department for guidance and encouragement during preparation of this book.

BONNIE A. CHARPENTIER
MICHAEL R. SEVENANTS
The Procter & Gamble Company
Cincinnati, OH 45224

December 30, 1987

Chapter 1

Physical Chemistry of Supercritical Fluids

A Tutorial

C. T. Lira

Department of Chemical Engineering, Michigan State University, East Lansing, MI 48824-1226

> Solubility behavior of compounds in supercritical fluids is interpreted using a simplified solution model. Solid-fluid and liquid-fluid phase equilibria data are presented and discussions of the phase behavior are provided. Phase transitions and transport properties are briefly discussed.

Critical phenomena have fascinated many researchers. Early solubility experiments with fluids above their critical points were published in 1879 (1). The simple cubic equation of state that van der Waals brilliantly conceived in 1873 has been shown to be capable of qualitatively describing virtually every type of phase behavior that has been experimentally observed (2-3), yet phenomena in high pressure fluids are still fascinating to researchers over a century later. Although many advances have been made in understanding of high pressure behavior, even qualitative prediction of binary phase behavior is often difficult without some experimental data. This paper explores the physical chemistry of fluids near their critical points. Although the term supercritical denotes conditions above critical temperature and pressure, interesting behavior occurs throughout the critical region. For this discussion, the term supercritical will refer to conditions above the critical temperature and near the critical pressure.

This discussion is prepared to provide introductory information for researchers who are not actively involved in supercritical fluid research, but who have interest in following developments in the field. While it is not possible in one short chapter to provide a summary of the work of the last century, several review papers and symposia collections have been published recently which provide excellent introductory material (4-9). In addition, McHugh and Krukonis have recently published an outstanding survey book of supercritical fluid studies (10).

Through the last century experimental data have been accumulated for a great many systems, but most data are for systems with critical points which don't differ greatly. Much current interest deals with systems where the critical properties and molecular interactions of the pure components differ greatly. Because of large differences in molecular interactions, the phase behavior is substantially different from that usually presented in textbooks. Although much has been learned, these molecularly asymmetric systems are still not understood well. Development of supercritical processes will evolve as the fundamental data base grows and researchers gain experience in working with these asymmetric systems.

Chemical processing techniques evolve over periods of many years and it seems unlikely that supercritical processes will rapidly replace distillation, liquid-liquid extraction, or adsorption in most processes. These unit operations are highly developed and many fundamental relationships and empirical correlations are used in process design. Further, each process is selected based on the properties of the mixture to be processed. The recent growth in availability of high pressure equipment has greatly facilitated an expansion in the number of fundamental supercritical studies, and the rate at which these studies contribute to our understanding should continue to accelerate. Thus, as we learn more, supercritical operations will become increasingly important alternatives in processing, separation, and analytical techniques.

Figure 1 illustrates a possible flowsheet for design of an arbitrary chemical process. After characterizing the materials in step 2, experience in working with many different processes usually permits selection of a unit operation (e.g. distillation or liquid-liquid extraction) without further modeling or design of the process. Because supercritical processes are not understood well, comparison of these operations with supercritical processes is more complicated. Further analysis of a supercritical process is necessary before a comparison may be made.

Steps 4 and 6 of Figure 1 ultimately determine the success of a supercritical process. However, without steps 3 and 5 most optimization must follow experimental step 7 which leads to slow and expensive process development. Steps 3 and 5 are well characterized for most processes, and although the design often involves empirical knowledge, we have enough fundamental understanding to optimize the process conditions. As we develop understanding of the relationships between steps 3-5 for supercritical systems, the development of an increasing number of successful high pressure processes will emerge due to improved process design at step 6. Steps 1-3 are dependent on the specific materials. The following discussion focuses on steps 4 and 5 to provide information to the majority of readers who wish to develop an understanding of the types of processing which may be achieved and the effects of temperature and pressure.

The application of supercritical processes will always be intimately coupled with the high pressure phase behavior and the physical chemistry of the system. Supercritical fluids have received much attention because the solubility of a solute is readily influenced by small variations in pressure or temperature. In both cases, the density of the supercritical fluid also changes. The solubility of a given compound may be influenced in several ways. Both vapor pressure and intermolecular forces determine solubilities. The following discussion should be helpful in understanding the effects.

Solubility of Solids in Supercritical Fluids

The simplest type of phase behavior to understand is the solubility of a solid solute, such as naphthalene, in a supercritical fluid. When the solute is a crystalline solid, the solid phase may be assumed to be pure and only the supercritical phase is a mixture. Imagine solid naphthalene in a closed vessel under one atmosphere of carbon dioxide at 40°C. The reduced temperature and reduced density of CO_2 are 1.03 and 3.7×10^{-3} respectively. At this pressure, the gas phase is ideal and the naphthalene solubility is determined by its vapor pressure. As the container volume is decreased isothermally, the solubility initially decreases when the gas phase is still nearly ideal. As the pressure is increased further, however, the gas phase density becomes increasingly nonideal and approaches the mixture critical density (near the critical density of CO_2 because the gas phase is still mostly CO_2). The reduced density of CO_2 increases rapidly near the critical region as shown in Figure 2. The solvent power of CO_2 is related to the density which leads to a rapid solubility increase. A brief description of intermolecular interactions is helpful in understanding this behavior.

For this discussion, we will consider a solute molecule with a dipole moment and then generalize the results. Imagine a molecule with a dipole moment at infinite dilution in a nonpolar supercritical fluid. The dipole's field induces a response of the polarizable fluid which results in a net attractive force. The polarizable solute reacts to the fluid's induced field and further increases the attractive force. A simplified mathematical model of these interactions is helpful in understanding the reasons for solubility enhancement and the density effect.

Consider a spherical point dipole solute molecule with a dipole moment, $\vec{\mu}$, at infinite dilution in a spherical container of supercritical fluid. With a continuum assumption, the fluid's electrical properties may be represented by a homogeneous dielectric constant, ϵ. The inhomogeneous field of the dipole polarizes the fluid which reacts and gives rise to a field, $\vec{R}$, at the dipole. $\vec{R}$ will be proportional to $\vec{\mu}$ as long as no saturation effects occur,

$$\vec{R} = f\vec{\mu} \qquad (1)$$

where f is called the factor of the reaction field. For our case a mathematical analysis provides

$$f = \frac{1}{a^3} \frac{2(\epsilon-1)}{(2\epsilon+1)} \qquad (2)$$

where
a = radius of point dipole
ϵ = dielectric constant of fluid

The interaction energy of a spherical nonpolarizable point dipole in its own reactive field is then

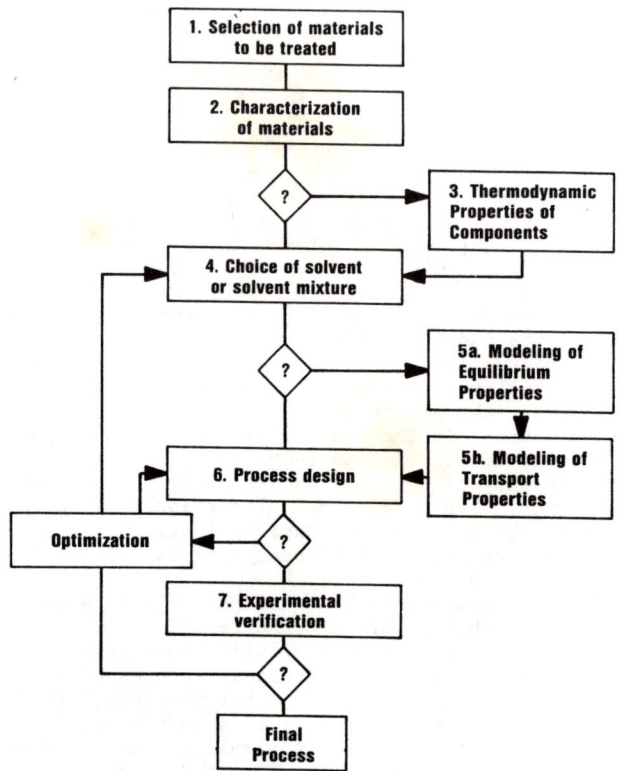

Figure 1. Scientific development of a chemical process.

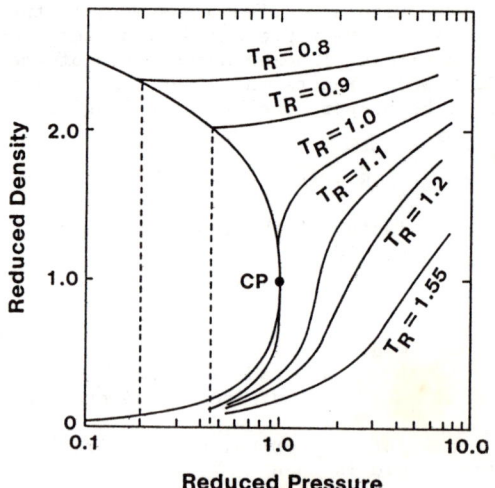

Figure 2. Reduced densities in the critical region along isotherms. (Reproduced with permission from Ref. 4. Copyright 1983. D. Reidel.)

$$\Gamma = -\frac{1}{2}\vec{\mu}\cdot\vec{R} = -\frac{1}{2}f\mu^2 = -\frac{\mu^2(\epsilon-1)}{a^3(2\epsilon+1)} \quad (3)$$

In a more realistic case, with a polarizable dipole, the result becomes

$$\Gamma = -\frac{1}{2}\frac{f\mu^2}{1-f\alpha} \quad (4)$$

where α = polarizability of point dipole

Note that this interaction energy is even more favorable since $(1-f\alpha)<1$. Derivations of the results presented here are discussed by Bottcher (11). The dielectric constant of a fluid may be represented by the Clausius-Mossotti function (12-13).

$$\frac{\epsilon-1}{\epsilon+2}\hat{V} = \frac{4\pi}{3}N_A\left[\alpha + \frac{\mu^2}{3kT}\right] \quad (5)$$

where $\hat{V}$ = molar volume
N_A = Avagadro's number
α = fluid polarizability
μ = fluid dipole moment

Equation 5 neglects interactions between quadrupoles, induced dipoles, and multiple bodies. An increase in temperature at constant density decreases the contribution of the fluid's dipole to the dielectric constant due to increased random motion. The polarizability is independent of temperature, although the polarizability of CO_2 reportedly decreases by 10% under a pressure of 100 to 200 atm (14-15). Dielectric constant measurements for CO_2, which has no dipole moment, show the Clausius-Mossotti function has a slight temperature and density dependence. The Clausius-Mossotti function is written

$$\frac{(\epsilon-1)}{(\epsilon+2)}\hat{V} = A_\epsilon + B_\epsilon/\hat{V} + C_\epsilon/\hat{V}^2 \quad (6)$$

where A_ϵ, B_ϵ, and C_ϵ are the dielectric virial coefficients. The dielectric virial coefficients for CO_2 at 323K are A_ϵ = 7.350 cm^3/mol, B_ϵ = 50.7 cm^6/mol^2, C_ϵ = -2515 cm^9/mol^3 (16).

From these data, we may use our simple dipole model to calculate the density dependence of the interaction energy. For the simple case of the non-polarizable point dipole the interaction energy of Equation 3 varies almost linearly with CO_2 density as shown is Figure 3. The critical density of CO_2 is 10.64 mol/l.

The simple model of Equation 3 considers a point dipole immersed in a fluid rather than a nonpolar, polarizable molecule. In the latter case, where the interactions result in dispersion forces rather than dipole-induced forces, the functionality of the forces permits extrapolation of the results from above. Since intermolecular pair potentials for dispersion forces and dipole-induced forces both vary as $1/r^6$, where r is the intermolecular distance, we expect both forces to scale the same with volume dependence, although the magnitudes of the forces will be different. In fact, for most situations, dispersion forces are much larger than dipole-induced forces (17).

The interaction energy for a nonpolar, spherical solute molecule (2) of radius a, surrounded by nonpolar solvent molecules (1) is ([18])

$$\Gamma = -\frac{3}{4} \frac{h\alpha_2}{a^3} \left[\frac{\nu_1 \nu_2}{\nu_1 + \nu_2} \right] \left[\frac{n^2-1}{2n^2+1} \right] \tag{7}$$

where h = Planck's constant
 ν = effective adsorption frequency
 n = refractive index of fluid

The density dependence of the refractive index is given by the Lorentz-Lorenz function

$$\frac{n^2-1}{n^2+2} \hat{V} = A_R + B_R/\hat{V} \tag{8}$$

with A_R= 6.658 cm³/mol, B_R= 3.3 cm⁶/mol² for CO_2 at 323K ([13]). Using this relationship, the interaction energy of Equation 7 is nearly linear with CO_2 density as shown in Figure 3. Relationships between the refractive index and the dielectric constant are discussed by Smyth ([19]).

Although density dependence of solvent power is described by Equations 3 and 7, a knowledge of solubility dependence is more useful. Smith and Walkley ([20]) compare solubility data for three different solutes in a variety of liquid organic solvents and show, for nonpolar systems, that the logarithm of solubility is related to the solvent strength in the dilute region at constant temperature. This result may be inferred from Figure 4 which has been reproduced from their work. Note that the polar solvents' behaviors deviate slightly from the behaviors of the other solvents. The solubilities of solid compounds in supercritical fluids also follow this behavior in the dilute region, as shown in Figure 5. (It will be shown later that a portion of this linear region is due to a cancellation of effects rather than purely attractive forces.) Although the curves exhibit a slight downward curvature, the solvent strength changes approximately linearly with density, as our model predicts. One more conclusion may be made from our model by examining Equations 3 and 7. Since the solubility depends on solvent strength, which may be related to the solvent's dielectric constant and refractive index, the models predict that a given compound will be more soluble in a supercritical fluid with a larger dielectric constant or refractive index at the same density and temperature. This behavior is usually true although repulsive forces, polarity, and acid-base effects are often important also.

These simple models permit a qualitative understanding of several factors influencing solubility, but cannot quantitatively predict solubility in supercritical fluids. Phase equilibrium calculations which are used to determine solubilities must consider energetic effects in all coexisting phases. Solubilities are determined by a minimum in Gibbs' energy when all phases are considered. Although we have discussed solubility behavior for a given compound, solubility comparisons between different compounds may not be made solely on these models of the supercritical phase. Also, these models will not predict the correct behavior at very

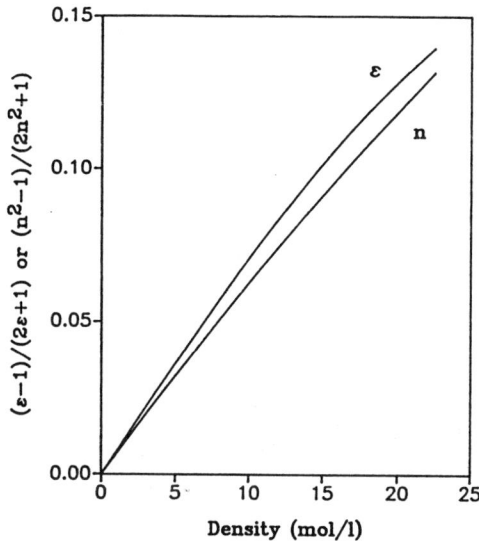

Figure 3. Density dependent portions of interaction energies as modeled by the dielectric constant and refractive index for CO_2.

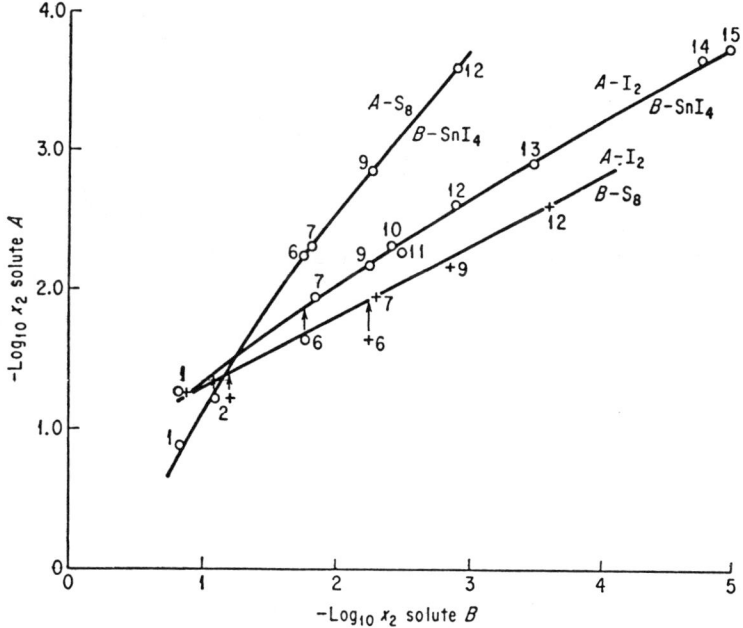

Figure 4. Solubility behavior of S_8, I_2, and SnI_4 in various solvents: 1,CS_2; 2,$CHBr_3$; 6,$CHCl_3$; 7,CCl_4; 9,$n-C_7H_{16}$; 10,$SiCl_4$; 11,$i-C_8H_{18}$; 12,CCl_2FCClF_2; 13,$c-C_4Cl_2F_4$; 14,$c-C_6F_{11}CF_3$; 15,C_7F_{16}. (Reproduced with permission from Ref. 20. Copyright 1960. The Royal Society of Chemistry.)

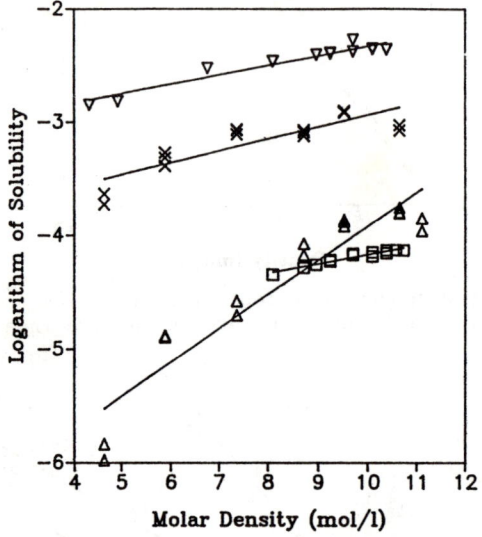

Figure 5. Solubility behavior of several solid compounds in supercritical SF_6 at 323 K: ▽ - naphthalene, X - dibenzofuran, △ - triphenylmethane, □ - acridine (Data from Ref. 21.)

high densities, because repulsive forces and multiple-body interactions, which we have not modeled, are important. In addition, the extrapolation to low pressures will be incorrect because the intermolecular forces upon which the model is based become negligible and solubilities are determined by the vapor pressure of the solute. The models are only meant to represent density effects in the intermediate region where the attractive forces are density dependent.

The continuum models of Equations 3 and 7 are inadequate for polar solvents partly due to the difficulty of describing the local field acting on each solvent molecule and the configuration of solvent molecules around the solute (22-23). Because of complicated behavior, empirical correlations are generally used to characterize solvent-solute interactions (24). Unfortunately, translations of empirical correlations into fundamental physical models are difficult. Empirical solvent-solute parameter studies in supercritical fluids are presented by Hyatt (25), Sigman, et al., (26), and Smith, et al. (27-28). Smith, et al., have investigated the π^* parameter (see Ref. 24) by observing the shift in the maximum UV adsorption wavelength of 2-nitroanisole in several supercritical solvents. Wavelength shifts are characteristic of changes in the solvent spheres (cybotactic region) surrounding the 2-nitroanisole molecules. For CO_2 the wavelength shifts may be related to Equation 7, although the relationship appears to be complex (27). Also, solubilities in supercritical fluids do not correlate directly with the π^* parameter (28).

Quantitative solubility calculations are usually performed using equation of state methods (10). Kim, et al., (29), discuss solubility behavior in the immediate vicinity of the critical point using macroscopic thermodynamic properties.

<u>The Enhancement Factor</u>

Characterization of the solvent power of supercritical fluids is often quantified by the enhancement factor. The enhancement factor is the ratio of the actual solubility to the ideal gas solubility at the same temperature

$$E = \frac{y_{actual}}{P^S/P} \qquad (9)$$

where P^S is the solid's vapor pressure. Note that $P^S/P = y_{ideal}$ is the ideal gas solubility, where gas phase nonidealities and other pressure corrections are omitted. Enhancement factors of the order 10^5-10^7 are common. Since the solubility changes rapidly near the critical region, the enhancement factor changes rapidly also. To simplify the representation of the enhancement factor, consider the logarithm,

$$\log_{10} E = \log_{10}(y_{actual}) - \log_{10}(P^S/P) \qquad (10)$$

which provides a difference rather than a ratio for the measure of solubility enhancement. From our discussion above, we expect the logarithm of solubility be approximately linearly dependent on density in the critical region, therefore the logarithm of the enhancement factor should be approximately linearly dependent on density. Linear behavior is observed experimentally as shown in Figure 6a. Although the logarithm of the solubility has a slight downward curvature, the logarithm of pressure has a slight upward curvature along the isotherm and the effects cancel.

The solubility and the enhancement factor are dependent on the interactions in the supercritical phase, but are also dependent on the properties in the solid phase. Equating the fugacities of the solid compound in both phases, and using the convention that component 2 is the solid solute,

$$f_2^V = f_2^S \qquad (11)$$

or

$$y_2 \hat{\phi}_2 P = \phi_2^S P^S \exp\left(\frac{v^S(P-P^S)}{RT}\right) \qquad (12)$$

where $\hat{\phi}_2$ is the fugacity coefficient of the solute in the supercritical phase, ϕ_2^S is the fugacity coefficient of the pure solute vapor at the pure solid vapor pressure P^S, and v^S is the molar volume of the pure solid. Thus, the enhancement factor is

$$E = \frac{\phi_2^S \exp\left(\frac{v^S(P-P^S)}{RT}\right)}{\hat{\phi}_2} \qquad (13)$$

The numerator quantifies the effect of hydrostatic pressure on the fugacity of the solid phase. The exponential term is known as the Poynting correction (17). The denominator quantifies the fluid phase intermolecular interactions and density effects. Note that the enhancement factor is dependent on the solid volume as well as the interactions in the supercritical fluid. A solute with a large solid molar volume will have a larger enhancement factor than a solute with a smaller solid molar volume at the same temperature and pressure when the interactions in the supercritical phase are identical. To further understand the molecular interactions in supercritical fluids, it is interesting to decompose the enhancement factor into these two effects. We may define a fluid enhancement factor, E_F, and a Poynting enhancement factor, E_P,

$$E = E_F E_P \qquad (14)$$

where $E_F = \frac{1}{\hat{\phi}_2}$ and $E_P = \phi_2^S \exp\left(\frac{v^S(P-P^S)}{RT}\right)$

When solid molar volumes are known, the Poynting correction effect may be divided out of the enhancement factor. The results are shown in Figure 6b for the data shown in Figure 6a. For all these solutes, the density of the supercritical phase is accurately approximated as the density of pure SF_6 since the solubilities are below $y = .004$. Along isotherms above the critical density, the fluid density increases slowly with large increases in pressure (review Figure 2). E_P increases rapidly in this region due to pressure changes (Equation 14) and E_F decreases rapidly. If the pressure is increased further, a maximum in the solubility may occur if the solute is squeezed out of solution by repulsive forces. Kurnik and Reid discuss this effect for naphthalene in ethylene (30). The solubility begins to decrease when the fluid enhancement, E_F, begins to decrease faster than the Poynting enhancement increases. Note in Figure 2 that the fluid density diverges from the ideal gas density as the phase envelope is approached along an isotherm from low pressure, but soon after the critical pressure is passed, the isotherm turns and begins to approach the ideal gas isotherm again. In the latter region, the repulsive forces are large. The simple models of Equations 3 and 7 are not valid in this region; Figure 5 remains linear in this region due to a cancellation of effects.

Melting of Solids in Supercritical Fluids

While the study of crystalline solids in supercritical fluids is relatively elementary, researchers must be cognizant that under appropriate conditions, solids melt substantially below their normal melting points. Figure 7 illustrates the phase behavior in a solid-fluid system. From a cursory inspection of the schematic, the diagram is more complicated than might be expected because the system is only a solid-fluid system in a certain region of the phase diagram. On the left hand portion of the figure, the curve labeled L_1V represents the pure vapor pressure of the lighter component (supercritical fluid). The pure component properties of the heavy component (solute) are denoted on the right hand side of the figure by the curves S_2V (sublimation), L_2V (vaporization), and S_2L (pure melting curve). All other curves on the figure denote phase behavior which occurs in binary systems. The region within the boundary of lines S_2LV, L=V, and L_2V on the right side of the figure represents an area where the solid has been melted and liquid-fluid equilibrium exists. The point K_2 is called the upper critical end point, and may be substantially below the pure solid melting point. Figure 8 illustrates the S_2LV lines for several solutes in ethane (31-33). Liquid-fluid behavior exists above the curves -- solid-fluid behavior exists below the curves.

Another region of liquid-fluid behavior exists near the critical point of the pure supercritical fluid. This behavior occurs within the region bounded by S_2LV, K_1, and L=V on the left of Figure 7. This region is usually very narrow, and therefore most published data are not in this region. Careful measurements of van Gunst, et al. (34), locate K_1 1.5 degrees Celsius and 1.2 atm above the critical point of pure ethylene for the system ethylene-naphthalene. For systems where the solid solubility is lower than naphthalene, this region is even smaller and closer to the pure solvent critical point.

While the melting point depressions for binary systems may not seem to be of great concern because most studies are performed near the solvent critical point, in ternary systems with two solids and one supercritical fluid the melting points may be lowered significantly. For example, the phase behavior of two solids in contact with a supercritical fluid is shown in Figure 9. The dotted lines represent the phase behavior in the pure systems and the binary systems discussed above. New phase boundaries in the ternary system are indicated by the solid lines. Note that the

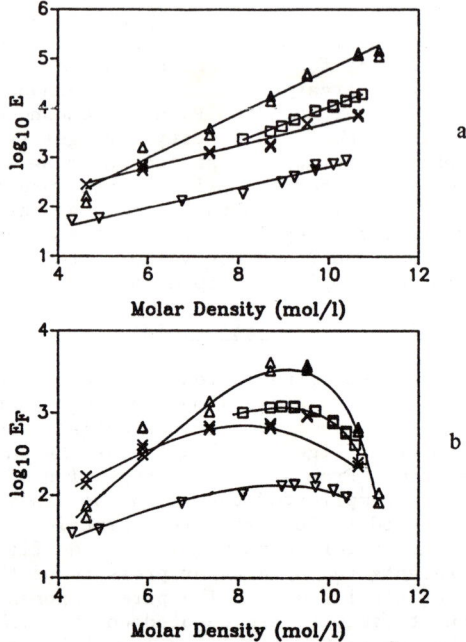

Figure 6. Behavior of enhancement factors, E, and fluid enhancement factors, E_F, for several solid compounds in supercritical SF_6 at 323 K: ∇ - naphthalene, X - dibenzofuran, △ - triphenylmethane, □ - acridine (Data from Ref. 21.)
(a) Linear behavior. (b) Poynting correction effect divided out of the enhancement factor.

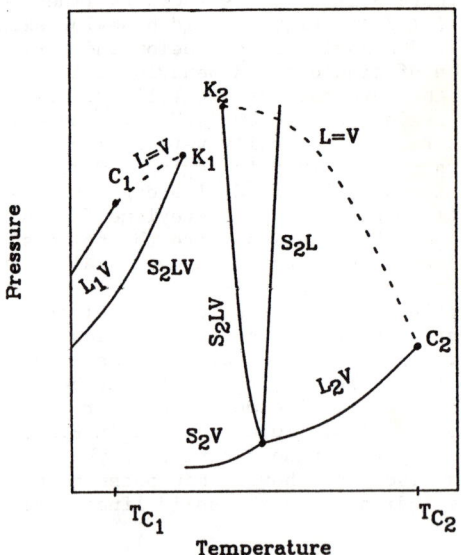

Figure 7. Typical phase behavior for binary systems with very large differences in pure component properties.

1. LIRA   Physical Chemistry of Supercritical Fluids

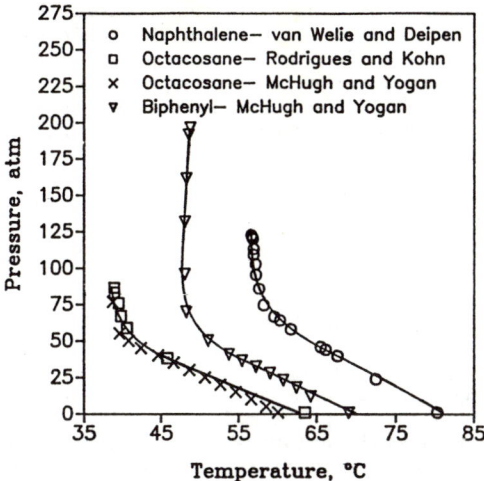

Figure 8. S_2LV lines for several compounds in supercritical ethane. (Data from Ref. 31-33).

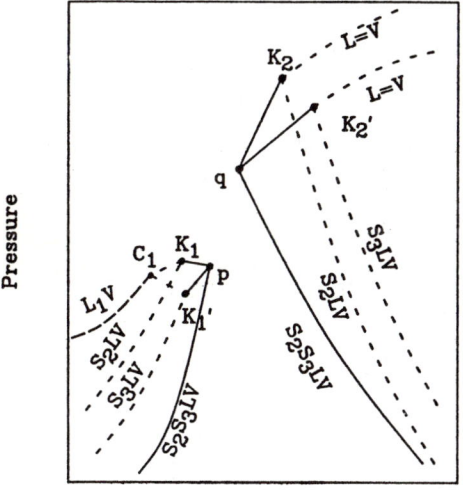

Figure 9. Typical phase behavior for ternary systems with very large differences in pure component properties. (Line segments: q-K_2, S_2+ (L=V); q-K_2', S_3+ (L=V); p-K_1, S_2+ (L=V); p-K_1', S_3+ (L=V)).

region of solid-fluid behavior is decreased further and the new critical end points (given by p and q) are much closer. As an example of the dramatic melting point depressions which may occur in such systems, the ethylene-naphthalene-hexachloroethane system is illustrated in Figure 10. The melting point of the pure compounds at atmospheric pressure are: naphthalene, 80.1°C; hexachloroethane, ≈190°C. The eutectic melting temperature of the solid mixture at atmospheric pressure is 56.6°C, and yet under 60 atm of ethylene, the solid mixture melts just above 25°C. Also note that point p extends several degrees beyond the critical temperature of ethylene. The liquid mixture contains dissolved ethylene. Very few experimental studies of this type exist in the literature; More studies would be helpful in developing predictive modeling.

Solid Solubilities in Multicomponent Systems

Solute/Cosolvent/Solvent Systems. Solubilities of solids may be modified by adding a small concentration of a nonpolar hydrocarbon (e.g. propane, octane) or a polar molecule (e.g. acetone, methanol). CO_2 has a small polarizability and no dipole moment, so additives increase the polarizability of the solvent (i.e. the refractive index of Equation 7) and the dielectric constant (Equation 3). Polar cosolvent molecules also interact with functional groups on the solutes. Cosolvents may increase solubilities up to an order of magnitude although the enhancement is dependent on cosolvent concentration. Unfortunately, relatively few fundamental cosolvent studies are published at this time.

Among the cosolvents studied, methanol and acetone have received the greatest interest ([36-38]). Methanol may act as either a Lewis acid or a Lewis base while acetone is a weaker Lewis base and very slightly acidic ([39]). The dipole moment of acetone is 2.88 Debeye compared to 1.7 Debeye for methanol. Based on these properties, Walsh, et al., ([40]), interpret the data of Van Alsten ([37]) and Schmitt ([38]) and present liquid phase IR measurements which show Lewis acid-base interactions in the systems methanol/acridine and acetone/benzoic acid. Supercritical solubility data of Dobbs, et al., ([36]), exhibit trends which indicate the importance of acid-base interactions. Van Alsten and Schmitt present data which show that acid-base interactions are a secondary cosolvent effect superimposed on a primary effect determined by cosolvent concentration.

Acid-base interactions may also occur between the cosolvent and the supercritical fluid. Hyatt ([25]) finds the basicity of CO_2 in the ether to ethyl acetate range. When interpreting Van Alsten's data, Walsh, et al., consider that CO_2 as a Lewis base, may interact with methanol and compete with acetone. Currently, very few data are available for study of these effects.

Yonker, et al., ([27]), present a study of CO_2/methanol mixtures at 50°C using UV absorbance maxima shifts for 2-nitroanisole. A 5.6 wt % methanol mixture exhibits a more pressure sensitive wavelength shift than pure CO_2, which indicates a more sensitive solvent sphere, however, the absorbance maximum for a 9.5 wt % mixture is insensitive to pressure, which is interpreted as an indication that the solute/solvent sphere does not change with

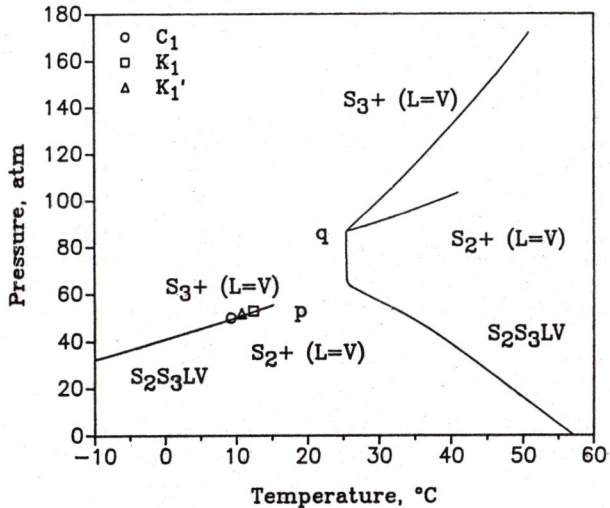

Figure 10. Phase boundaries for the system CO_2-naphthalene-hexachloroethane. (Data from Ref. 35.) (2-hexachloroethane, 3-naphthalene).

density. The relationships between these measurements and solubility behavior are not well understood (28).

Solubilities of nonfunctional compounds are also enhanced by cosolvents, but the enhancements are more dependent on the cosolvent concentration than the cosolvent functionality (37-38). Nonpolar, polarizable hydrocarbon cosolvents seem more effective than polar cosolvents for enhancing nonfunctional compound solubilities in CO_2 (36,38), although the differences are small. These differences are negligible in ethane (38), which is more polarizable than CO_2.

Cosolvent studies must be carefully performed. Addition of the cosolvent shifts the critical properties from the pure solvent critical properties. These conditions must be known to assure that the mixture remains homogeneous throughout the range of experiments. Also, the cosolvent may create significant melting point depression of solids. Dobbs, et al., (36), report that a 3.5 mol % methanol/CO_2 solvent melts resorcinol (T_m = 111°C) at 35°C and pressures between 100-350 bar.

Solute/Solute/Solvent Systems. Solute/solute/solvent studies are presented by Kurnik and Reid, (41), Kwiatkowski, et al., (42), and Gopal, et al., (43). Unfortunately, some studies have not proven clearly that solutes were always solids under experimental conditions. In general it is noted, for solid solutes, that the ternary solubility of a less soluble solid is greatly increased by the presence of a significantly more soluble solid (36). For example, in the naphthalene/phenanthrene/CO_2 system, naphthalene increases the solubility of phenanthrene and the ternary system selectivity is substantially below the selectivity predicted from the ratio of binary solubilities. Thus, substantially soluble solids act as cosolvents for less soluble compounds.

Solubility of Liquids in Supercritical Fluids

A solute that is a liquid at the temperature of a supercritical extraction has critical properties much closer to the supercritical fluid's critical properties than the solids discussed above. Since the mixture is somewhat less molecularly asymmetric, the critical temperature of the supercritical fluid lies above the melting temperature of the solute. As a result, the vaporization (L_2V) curve is generally the only pure solute property to be of concern on mixture P-T traces. Many literature references are available for mixtures that fall into this category, because these systems comprise the bulk of high pressure vapor-liquid equilibria research of the last century, however, most data are for hydrocarbon related systems and the current interest extends beyond these systems. A few cases of special interest will be mentioned and further information may be found in other references (4-10).

High pressure vapor-liquid behavior is typically classified into one of five basic types illustrated in Figure 11. (Classification numbering of the systems varies and one additional classification is sometimes added (8,10,17)). If the only area of interest is above the critical point of the lighter component, then the only temperatures of interest are above C_1. As an illustration of type III behavior, a P-x-y diagram of the ethylene-n-propanol

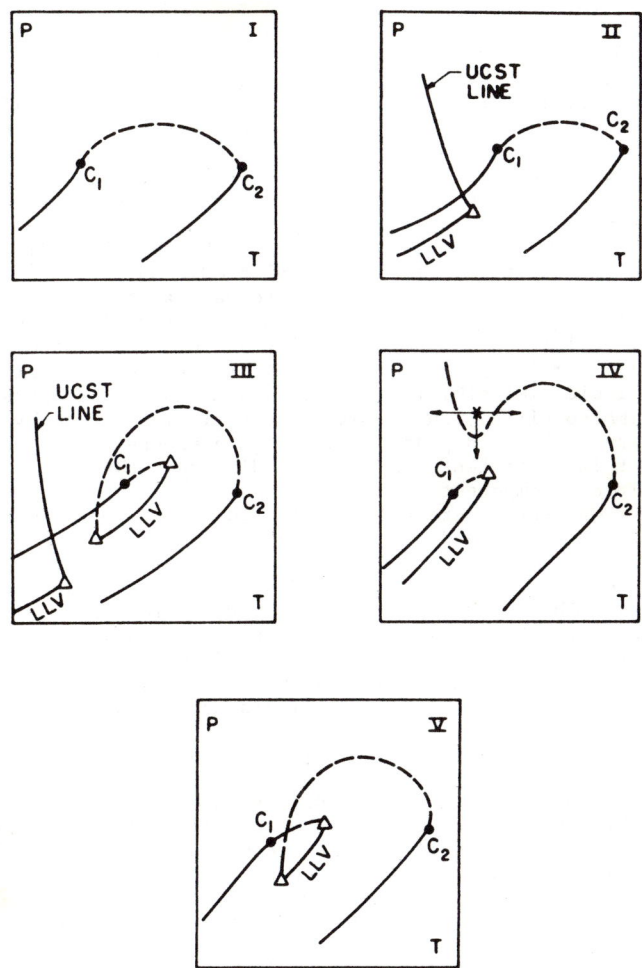

Figure 11. Five classes of binary phase behavior typically observed. (Reproduced with permission from Ref. 10. Copyright 1986 Butterworths.)

system is shown in Figure 12. Comparing this figure to Figure 11c, the temperature is above the critical temperature of the light component, but below the temperature of the critical endpoint denoted by the triangle. A wide range of interesting liquid-liquid-fluid behavior exists in these systems in subcritical regions also.

Compare the behavior of the ethylene-n-propanol system to a type I system, CO_2-hexane, shown in Figure 13. Note that in both cases, solubility of the supercritical component in the liquid phase increases rapidly as the pressure increases. This phenomenon can lead to substantial swelling of the liquid phase. The solubility of the heavy component in the lighter phase does not increase rapidly with an increase in pressure until the mixture critical point is approached or an additional phase is formed. The conditions of these occurrences may be substantially removed from the pure supercritical fluid critical conditions.

Interesting behavior occurs in ternary systems when two immiscible liquids are contacted with a supercritical fluid. The supercritical fluid is preferentially dissolved in the most chemically similar phase, effectively increasing the degree of liquid-liquid immiscibility (46). In addition, it is possible to create immiscibility from a homogeneous liquid mixture by contacting the liquid phase with a high pressure gas if one liquid component is sufficiently chemically different than the supercritical component.

Most current interest in liquid-fluid behavior focuses on the separation of dissolved organic species from water. Studies with both alcohols and organic acids are published. Results are usually summarized on ternary diagrams similar to diagrams used for liquid-liquid equilibria. The n-propanol-water-ethylene system is illustrated in Figure 14. The supercritical phase compositions are very nearly pure ethylene and cannot be discerned from this diagram. Note the significant solubility of the supercritical fluid in the liquid, and also the immiscibility effect.

Transport Properties of Supercritical Fluids

Transport properties, including diffusion coefficients and viscosities, undergo changes in the critical region. As mentioned in Figure 1, these properties are useful in optimizing supercritical processes and the use of these properties should become more important as supercritical fluid process calculations develop. This discussion is presented to stress the density dependence of these properties.

<u>Viscosity.</u> Viscosity measurements are available for a variety of fluids at high pressure (48). Several corresponding states methods have been developed using different characteristic viscosities. One of the easiest techniques to interpret for an introduction is the technique used by Comings, et al., (49). These researchers used a reduced viscosity based on the viscosity at P=0,
$$\eta^{\#} = \eta(T,P)/\eta(T,P=0).$$
The value of the viscosity at 1 atm is a satisfactory approximation for the zero pressure viscosity. As shown in Figure 15, the viscosity changes by a factor of 3 or 5 in the critical region,

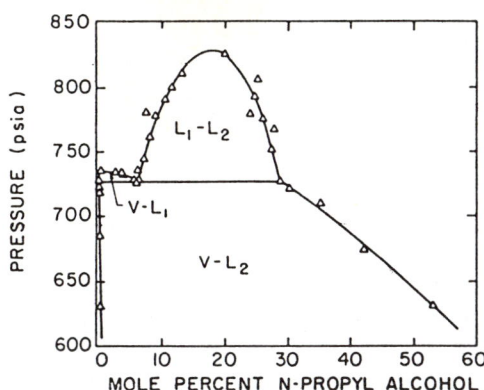

Figure 12. Binary phase behavior of ethylene-n-propanol at 14.5 °C, a type III system. (Reproduced with permission from Ref. 10. Copyright 1986 Butterworths.)

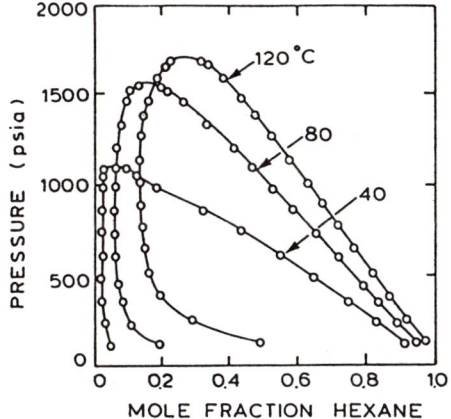

Figure 13. Binary phase behavior of CO_2-n-hexane, a type I system. (Reproduced with permission from Ref. 10. Copyright 1986 Butterworths.)

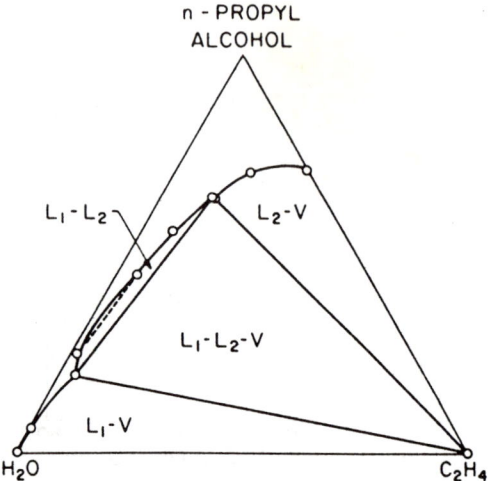

Figure 14. Ternary phase behavior of n-propanol-water-CO_2 at 15 °C and 750 psia. (Reproduced with permission from Ref. 10. Copyright 1986 Butterworths.)

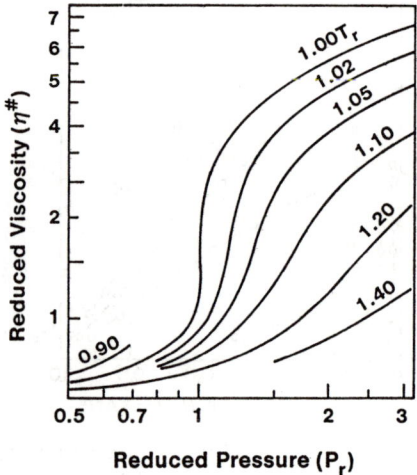

Figure 15. Reduced viscosity in the supercritical region. (Data from Ref. 49.)

rather than orders of magnitude. At higher reduced temperatures the pressure effect is less pronounced. Compare Figure 15 with Figure 2 to note that the viscosity and density changes are coupled.

Diffusion Coefficients. Self diffusion coefficients for CO_2 (50-52), ethylene (53), water (54), and methane (55) are presented in Figure 16. The critical densities of these fluids are 10.6, 7.8, 17.9, and 10.1 mol/l, respectively. Figure 16 is presented for illustrative purposes only and the references provide a discussion of theoretical considerations and mathematical relationships between density, viscosity, and diffusion.

Binary diffusion coefficients are available for a few mixtures. The diffusion values are between the solvent self-diffusion coefficients and normal liquid phase diffusion coefficients and the values are summarized in Table I.

Table I. Binary Diffusion Coefficients in Supercritical Fluids

Solvent (Ref.)	Solute	Temperature (K)	Density (mol/l)	Approximate Range $-\log_{10}(D_{12}[cm^2/s])$
CO_2 (56)	Benzene Phenol Naphthalene	313	7-18	3.5-4.0
CO_2 (57)	Benzene n-Propylbenzene 1,3,5-Trimethyl-benzene	313	7-18	3.5-4.0
CO_2 (58)	Benzoic acid 2-Naphthol	308-323	15-17	4.04-4.30
SF_6 (58)	Benzoic acid Naphthalene	318-338	7.5-10	3.8-4.08
2,3-Dimethylbutane (59)	Benzene Toluene Naphthalene Phenanthrene	523-548	4.0-5.3	3.4-3.6

In general, mass transfer is not limited by diffusion rates in the supercritical phase. In fact, significant buoyancy effects in supercritical fluids enhance mass transfer rates with convective mixing (58). Usually, mass transfer limitations will occur in either a liquid or solid phase.

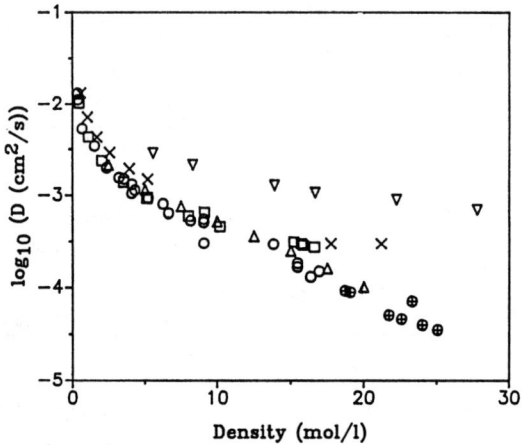

Figure 16. Self diffusion coefficients for several fluids: O - CO_2, (313-318K, 1.03-1.05 T_r), (50); ⊕ - CO_2, (323K, 1.06 T_r), (51); □ - CO_2, (308K, 1.01 T_r), (52); Δ - C_2H_4, (323K, 1.14 T_r), (53); ∇ - H_2O, (673K, 1.04 T_r), (54); X - CH_4, (198K, 1.04 T_r), (54).

Conclusion

This chapter provides an introduction to supercritical fluid behavior. As a tutorial, qualitative fluid behavior is stressed rather than quantitative description. Solubilities in solid-fluid systems are interpreted with a simple fluid model. Enhancement factors are introduced to demonstrate the importance of repulsive forces. Intermolecular interactions in cosolvent systems are discussed. Liquid-fluid phase behavior and phase transitions in liquid-fluid and solid-fluid systems are briefly presented. Transport properties are briefly presented to stress their density dependence.

Many introductory topics have been omitted from this discussion, such as polymer behavior and mixture fractionation (see Ref. 10). Reactions in supercritical fluids are also discussed elsewhere (60). Interesting applications of supercritical processing of materials are published, including powder manufacturing (10,60) and drying of gels (62). Although the number of applications continues to grow, each process requires steps where an understanding of phase behavior facilitates process optimization. As an overview, this chapter should be helpful in understanding more detailed studies.

Literature Cited

1. Hannay, J.B.; Hogarth, J. Proc. R. Soc. London 1879, 29, 324.
2. Scott, R.L.; van Konynenburg Disc. Faraday Soc. 1970, 49, 87-97.
3. Scott, R.L. Ber. Bunsenge. Phys. Chem. 1972, 76, 296-307.
4. Paulaitis, M.E.; Krukonis, V.J.; Kurnik, R.T.; Reid, R.C. Rev. Chem. Eng. 1983, 1, 179-249.
5. McHugh, M.A. In "Recent Developments in Separation Science," Li, N.N. and Calo, J.M., Eds.; CRC Press: Boca Raton, FL, 1986; Vol. IX, 75-105.
6. Ber. Bunsenge. Phys. Chem. 1984, 88.
7. "Supercritical Fluid Technology," Penninger, J.M.L.; Radosz, M.; McHugh, M.A.; Krukonis, V.J., Eds.; Process Technology Proceedings, 3; Elsevier: Amsterdam, 1985.
8. "Chemical Engineering at Supercritical Conditions," Paulaitis, M.E., Penninger, J.M.L., Gray, R.D. and Davidson, P., Eds.; Ann Arbor Science: Ann Arbor, MI, 1983.
9. "Supercritical Fluids: Chemical and Engineering Principles and Applications," Squires, T.G. and Paulaitis, M.E., Eds.; ACS Symposium Series, #329, American Chemical Society: Washington, DC, 1987.
10. McHugh, M.A.; Krukonis, V.J. "Supercritical Fluid Extraction: Principles and Practice," Butterworths: Boston, MA, 1986.
11. Bottcher, C.J., "Theory of Electric Polarisation," Elsevier: Amsterdam, 1952, 63-74, 133-139.
12. Vaughn, W.E.; Smyth, C.P.; Powles, J.G. In "Physical Methods of Chemistry," Weissberger, A. and Rossiter, B.W., Eds.; Wiley-Interscience: New York, 1972; Vol. I, pt. IV, 351-396.
13. St-Arnaud, J.M.; Bose, T.K. J. Chem. Phys. 1979, 71, 4951.
14. Bottcher, C.J. Physica 1942, 9, 937, 945.

15. deGroot, S.R.; ten Seldam, C.A. Physica 1947, 13, 47.
16. Bose, T.K.; Cole, R.H. J. Chem. Phys. 1970, 52, 140.
17. Prauznitz, J.M.; Lichtenthaler, R.N.; de Azevedo, E.G. "Molecular Thermodynamics of Fluid-Phase Equilibria," Prentice-Hall: Englewood, Cliffs, NJ, 1986; 2nd ed.; pp 48-90.
18. Linder, B. J. Chem. Phys. 1962, 37, 963.
19. Smyth, C.P. In "Physical Methods of Chemistry," Weissberger, A. and Rossiter, B.W. Eds.; Wiley-Interscience: New York, 1972; Vol. I, pt. IV, Chapter VI.
20. Smith, E.B.; Walkley, J. Trans. Faraday Soc. 1960, 56, 219.
21. Hansen, P.C., Ph.D. Thesis, University of Illinois-Urbana, 1985.
22. Onsanger, L. J. Am. Chem. Soc. 1936, 58, 1486.
23. Van Vleck, J.H. Mol. Phys. 1972, 24, 341.
24. Abboud, J.M.; Guihenuef, G.; Essfar, M.; Taft, R.W.; Kamlet, M.J. J. Phys. Chem. 1984, 88, 4414.
25. Hyatt, J.A. J. Org. Chem. 1984, 49, 5097.
26. Sigman, M.E.; Lindley, S.M.; Leffler, J.E. J. Am. Chem. Soc. 1985, 107, 1471.
27. Yonker, C.R.; Frye, S.L.; Kalkwarf, D.R.; Smith, R.D. J. Phys. Chem. 1986, 90, 3022.
28. Smith, R.D.; Frye, S.L.; Yonker, C.R.; Gale, R.W. J. Phys. Chem. 1987, 91, 3059.
29. Kim, S.; Wong, J.M.; Johnston, K.P. In "Supercritical Fluid Technology," Penninger, J.M.L.; Radosz, M.; McHugh, M.A.; Krukonis, V.J., Eds.; Process Technology Proceedings, 3; Elsevier: Amsterdam, 1985; pp. 45-66.
30. Kurnik, R.T.; Reid, R.C. AIChE J. 1981, 27, 861.
31. van Welie, G.S.A.; Diepen, G.A.M. J. Phys. Chem. 1963, 67, 755.
32. Rodrigues, A.B.; Kohn, J.P. J. Chem. Eng. Data 1967, 12, 191.
33. McHugh, M.A.; Yogan, T.J., J. Chem. Eng. Data, 1984, 29, 112.
34. van Gunst, C.A.; Scheffer, F.E.C.; Diepen, G.A.M. J. Phys. Chem. 1953, 57, 578.
35. van Gunst, C.A.; Scheffer, F.E.C.; Diepen, G.A.M. J. Phys. Chem. 1953, 57, 581.
36. Dobbs, J.M.; Wong, J.M.; Lahiere, R.J.; Johnston, K.P. Ind. Eng. Chem. Res. 1987, 26, 56-65.
37. Van Alsten, J.G., Ph.D. Thesis, University of Illinois-Urbana, 1986.
38. Schmitt, W.J., Ph.D. Thesis, Massachusetts Institute of Technology, 1984.
39. Kamlet, M.J.; Abboud, J.M.; Abraham, M.H.; Taft, R.W. J. Org. Chem. 1983, 48, 2877.
40. Walsh, J.M.; Ikonomou, G.D.; Donohue, M.D., Paper presented at the 1986 AIChE Meeting, Miami Beach, FL.
41. Kurnik, R.T.; Reid, R.C. Fluid Phase Equilibria 1982, 8, 93.
42. Kwiatkowski, J.; Lisicki, Z.; Majewski, W. Ber. Bunsenges Phys. Chem. 1984, 88, 865.
43. Gopal, J.S.; Holder, G.D.; Kosal, E. Ind. Eng. Chem. Process Des. Dev. 1985, 24, 697.
44. Todd, D.B.; Elgin, J.C. AIChE J., 1955, 1, 20.
45. Li, Y.-H.; Dillard, K.H.; Robinsin, R.L. J. Chem. Eng. Data 1981, 26, 53.

46. Yorizane, M.; Masuoka, H.; Ida, S.; Ikeda, T. J. Chem. Eng. Japan 1974, 7, 379.
47. Weinstock, J.J., Ph.D. Dissertation, Princeton University, 1952.
48. Reid, R.C.; Prausnitz, J.M.; Sherwood, T.K. "The Properties of Gases and Liquids," Third Edition, McGraw-Hill: New York, 1977.
49. Comings, E.W.; Mayland, B.J.; Egly, R.S. "The Viscosity of Gases at High Pressures," Engineering Experiment Station Bulletin, Series 354, Vol. 42, No. 15, University of Illinois, November 28, 1944.
50. Robb, W.L.; Drickamer, H.G. J. Chem. Phys. 1951, 19, 1504.
51. Timmerhaus, K.D.; Drickamer, H.G. J. Chem. Phys. 1952, 20, 981.
52. O'hern, Jr., H.A.; Martin, J.J. Ind. Eng. Chem., 1955, 47, 2081.
53. Baker, E.S.; Brown, D.R.; Jonas, J. J. Phys. Chem. 1984, 88, 5425.
54. Lamb, W.J.; Hoffman, G.A.; Jonas, J. J. Chem. Phys. 1981, 74, 6875.
55. Dawson, R.; Khoury, F.; Kobayashi, R. AIChE J. 1970, 16, 725.
56. Feist, R.; Schneider, G.M. Sep. Sci. Tech. 1982, 17, 261.
57. Swaid, Il; Schneider, G.M. Ber. Bunsenges. Phys. Chem. 1979, 83, 969.
58. Debenedetti, P.G.; Reid, R.C. AIChE J. 1986, 32, 2034.
59. Sun, C.K.J.; Chen, S.H. AIChE J. 1985, 31, 1904.
60. Subramanian, B.; McHugh, M.A., Ind. Eng. Chem. Process Des. Dev. 1986, 25, 1.
61. Matson, D.W.; Peterson, R.C.; Smith, R.D. Adv. Cer. Mat. 1986, 1, 242.
62. Laudise, R.A.; Johnson, D.W. J. Non-Cryst. Solids 1986, 79, 155.

RECEIVED November 11, 1987

Chapter 2

Processing with Supercritical Fluids

Overview and Applications

Val J. Krukonis

Phasex Corporation, 360 Merrimack Street, Lawrence, MA 01843

> When supercritical fluid extraction first came under investigation in the U.S. it received much research effort because of its potential for carrying out separations with low energy consumption. Such investigations received a great deal of publicity in trade and technical journals. Because the process development activities were mostly unsuccessful, interest in supercritical fluids waned; during the past three years, however, resurgent interest developed. A review of the events of the past ten years, namely, the events that created the initial interest in supercritical fluid extraction, the events that caused (temporarily) a decrease in interest in the technology, and the factors responsible for the current resurgent interest is developed. Two examples of supercritical fluid extraction applications that are of interest to food and flavor chemists, fish oils concentration and enzyme-catalyzed reactions, are described.

This chapter describes a recent and controversial period of supercritical fluid history, 1977-1987. An outline of the information to be covered in this chapter is given in Table I. The history of supercritical fluid solubility phenomena was summarized in an earlier paper (1). That paper reviewed the first literature report on the subject by Hannay and Hogarth in 1879 (2), the work of many researchers who investigated the phase behavior of various materials dissolved in supercritical fluids (3-5), and some process/product applications of supercritical fluid extraction (6-8). A quite detailed historical development, covering in depth the first score years after 1879, has been published elsewhere (9).

0097–6156/88/0366–0026$06.00/0
© 1988 American Chemical Society

Table I. PROCESSING WITH SUPERCRITICAL FLUIDS

OVERVIEW -- 1977-1987
- The Promises
- The Loss of Luster
- The Monotonic Recovery

APPLICATIONS
- Lipids (Fish oils concentration, cholesterol extraction from animal fats)
- Enzyme Reactions

Overview

In the early stages of process development in the U.S., many "promises" were made about the capabilities of, and applications for, supercritical fluid extraction. When the widely-touted applications did not materialize, as exemplified by commercialized processes, for example, the "luster" of supercritical fluid solvents rather quickly faded. It is informative to review some of the trade and association journal articles that were published during this period because of their influence on many researchers and companies in their future investigations of supercritical fluid extraction.

Some of the titles which appeared during the early period of investigation of supercritical fluid extraction in the U.S. included the following: "Critical fluids aim for a broad industrial role" (Chem. Eng., March 1979), "Gas solvents: about to blast off" (Bus. Week, July 1981), "Supercritical Fluids Promise Quick Extraction of Food Volatiles" (Food Devel., August 1981), "The promise for supercritical fluids" (January 1983), "Separation technology - keep an eye on supercritical fluid extraction" (Chem. Proc., January 1985), and "The magic of supercritical fluids" (Chem. Eng., February 1985). The articles contained pervasive promises, as reflected in some selected phrases from the articles, e.g., "tremendous dissolving capacity", "energy saver", "superior solvents", "significant alternative to distillation", "on the threshold of a new technology", "nothing less than a new unit operation", and "magic solvents". The lack of commercial success in the areas that had been suggested in the articles, for example, for replacing distillation in the separation of miscible liquids, resulted in a reversal of interest in supercritical fluids as extration solvents. The title of another article, "Supercritical fluids: still seeking acceptance" (Chem. Eng., February 1985), reflected the attitude that after almost ten years of effort it wasn't clear where supercritical fluids fit.

One explanation for the lack of success during this early period of activity resides in the then existence of unbridled enthusiasm about the potential for supercritical fluids. There were many misapplications, especially in the research effort that was directed to large volume chemicals separated by distillation. Perhaps in partial reconciliation of the misapplication of supercritical fluid extraction to separating large volume chemicals, it is important to recall that during the late 70's, which was coincident with rising interest in supercritical fluids in the U.S., energy conservation became an important consideration for new process development or old process improvement. In some cases supercritical fluid extraction was shown to be a lower energy consuming process than distillation, and it therefore was quite widely directed to solving some of the energy ills of the Chemical Process Industries. Energy savings alone, however, do not assure economic viability of any process.

Some of the process and applications development effort that was carried out during this period could have been predicted to be fruitless based upon then existing data. For example, much of the information necesary to evaluate the potential for supercritical fluid extraction, especially for separating many chemicals from water, was obtainable from a paper by Alfred Francis (10). His work was, when it was published in 1954 (and perhaps it is still today), the greatest single source of solubility and distribution data of compounds in liquid carbon dioxide. Because the progression from liquid to supercritical solvent properties is continuous, the liquid carbon dioxide data of Francis are valuable for estimating solubility and extraction behavior in the supercritical fluid regions as well.

As an example of the breadth and depth of Francis' studies one figure from reference 10 is reproduced in Figure 1; it shows many ternary phase diagrams of carbon dioxide (at conditions of 800 psi, 25°C). The top apex of each phase diagram represents carbon dioxide and the other two apices, respectively, two other compounds whose identities are obtained from a master table in the article. For example, in the top right ternary diagram (D4) shown in Figure 1, HD represents n-hexadecane and HN, hydrocinnamaldehyde, and from this phase diagram, much information can be obtained, e.g., the solubility of n-hexadecane in CO_2, the solubility of CO_2 in n-hexadecane, and similarly for hydrocinnamaldehyde. Additionally the concentrations of all components that result in complete miscibility or in three-component two-phase regions can be obtained. Francis reported on 464 ternary systems that included compounds ranging from simple hydrocarbons, alcohols, esters, and aldehydes to much more complex substituted aromatics and heterocyclics. The breadth and importance of his contributions cannot be overestimated, yet, apparently his work was quite frequently not consulted for its relevance; a number of research groups subsequently "rediscovered" that certain separations were not advantageously carried out by supercritical fluid extraction relative to processing by distillation.

The third entry in Table I is "The monotonic recovery", implying that the interest in supercritical fluid extraction is again increasing. Although the less than lustrous performance of supercritical fluids in energy reduction applications delayed or reduced other process development activity, currently there is increasing activity in supercritical fluid extraction. The resurgent interest resides in a number of factors, e.g.,

2. KRUKONIS   *Processing with Supercritical Fluids*   29

Figure 1. Ternary phase diagrams of carbon dioxide and other compounds at 800 psi and 25 °C. (Reproduced with permission from Ref. 10. Copyright 1954 J. Phys. Chem.)

increasing scrutiny of traditional solvents, increasing emphasis on pollution control, and increasing performance demands on products. These factors have motivated companies and government agency laboratories to evaluate alternative processes for solutions to specific separations or processing problems, and supercritical fluid extraction is one of the processes that are being applied and evaluated. Whereas in the early 80's the development of generic processes utilizing supercritical fluids for separating with low energy consumption a broad spectrum of products in the chemical process industries was a desired (and perhaps unachievable) goal, current successes are resulting from application of the technical and economic merits of supercritical fluids on a case by case basis. Although carbon dioxide can technically replace a variety of targeted solvents, i.e., carbon dioxide can "do" the extraction, economics must be evaluated early on in order to determine the commercial viability of supercritical fluid extraction. Furthermore, as is required for any realistic evaluation, all alternative separations processes and their capital costs and operating costs must be weighed in developing the optimum process to separate a specific mixture. Familiarity with the pros and cons of supercritical fluid extraction and with the reasons for the early "failures" has aided researchers in their selections of potentially viable applications.

Applications

A variety of applications demonstrate potential technical and economic viability, and a few are listed in Table II.

Table II. CURRENT SUPERCRITICAL FLUID EFFORT

- Recrystallization of Pharmaceuticals ([12], [13])
- Fractionation of Oils and Polymers ([14], [15])
- Extraction of Residual Solvents and Monomers from Polymers ([16])
- Polymerization in Supercritical Fluids ([17])
- Treatment of Liquid and Solid Wastes ([18], [19])
- Enzyme Reactions in Supercritcal Fluids ([20-22])
- Extraction of Flavors, Aromas, and Colorants ([1], [23], [24])
- Concentration of Eicosapentanoic Acid from Fish Oils ([25-29])
- Extraction of Cholesterol from Butter, Lard, Tallow & Eggs ([30])
- Decaffeination of Coffee ([31])
- Extraction of Hops ([32])

Not every pharmaceutical will eventually be comminuted by supercritical fluid nucleation, not every polymer processed for molecular weight control by supercritical fluid extraction, not every flavor concentrated by supercritical fluid extraction; but some will be. Two applications listed in the table are already in commercial production, and several are in advanced pilot plant development and test market evaluation. Hops extraction is being carried out by Pfizer, Inc. in its plant in Sydney, NE ([33]), and General Foods Corporation has constructed a coffee decaffeination plant in Houston, TX which will come on stream in late 1987 (34). Information about these plants is becoming quite widely known, and the fact that these plants exist is also rekindling interest in supercritical fluid extraction; it is obvious now that supercritical fluid extraction "works" somewhere and that it can be scaled. In Europe several supercritical fluid extraction plants processing coffee, hops, tea, and some selected spices have been operating for many years.

Cholesterol extraction studies listed in Tables I and II have progressed from the initial scale to an optimization phase at the laboratories of the University of Wisconsin and Phasex Corporation (35). Whereas all the applications listed in Table II are interesting for discussion, fish oils extraction and enzyme reactions were selected for their newness and/or novelty for discussion here. Fish oil extraction with supercritical fluids exhibits the potential to become the preferred separations process if high concentration levels of eicosapentanoic acid are required, and it is being investigated by many workers. The subject of enzyme reactions in supercritical fluids is, at present for the most part, an interesting academic pursuit but it exemplifies the breadth of application of supercritical fluid extraction.

Fish Oils Concentration

Because evidence has accumulated that suggests certain polyunsaturated fatty acids found in fish oils may have beneficial effects for the cardiovascular system, eicosapentanoic acid (EPA) has become the subject of intense study during the past few years. Several methods that exhibit the technical capabilities to concentrate EPA suffer from certain limitations in their operation and scale up. High vacuum and molecular distillation which can separate many low vapor pressure components require operation at quite high temperature levels to achieve any reasonable production rates, and the high temperature level can cause degradative reaction of the unsaturated components in fish oils. Chromatographic techniques such as HPLC, also demonstrating the technical ability to produce a high purity EPA "peak", suffer because of scale up limitations. As related earlier, in a case-by-case evaluation supercritical fluids extraction can emerge as a technically and economically superior process especially when product performance demands place limitations upon conventional processing techniques. In the case of producing a high concentration EPA product, supercritical fluid fractionation is being investigated because it can achieve a 90+% concentration, simultaneously a 90+% yield, and it can readily be scaled to production levels.

Triglycerides cannot be concentrated to any significant degree by supercritical fluid fractionation; this inability to concentrate EPA (other than by a few per cent (26)) is a result of the random association of fatty acid chains on a glycerol backbone. On any single glycerol moiety there might be, for example, a C18:1, a C20:5, and a C22:1 chain, and it is impossible to "extract" C20:5 from the triglyceride. Saponification and esterification to the ethyl esters enables a supercritical fluid to separate the assembly of individual esters according to the distribution coefficients of each. The selectivity of carbon dioxide for one ester over another, i.e., the ratio of the respective distribution coefficients, can be tailored by virtue of operating pressure.

The first report on the supercritical fluid fractionation of fish oils appeared in April 1984. Eisenbach (25), working wih ethyl esters of codfish oil, showed that by operating in the retrograde condensation region, i.e., that region of supercritical fluid phase space where an isobaric increase in temperature causes a decrease in the solubility of a dissolved compound, he could isolate a fraction containing 90+% C20 esters. The fraction also contained some C18's and C22's, but 90+% of the fraction was composed of C20's. The supercritical fluid fractionation was carried out using an internal reflux brought about by a "hot finger", i.e., a heated zone located physically above the extraction zone, which causes a partial liquefaction of the components that are dissolved in the gas stream leaving the extraction zone. The liquid ester reflux that falls downward simulates somewhat a continuous countercurrent extraction and brings about a concentration maximum in C20 esters and in EPA during the course of the extraction. Incidentally, this report of 90% concentration of C20 esters was later misinterpreted to be a 90% concentration of EPA in the product (36), and this same misinterpretation was repeated later (37). Many misperceptions prevail still today about the separations that are achievable (or not achievable) by supercritical fluid extraction, and thus, the confusion on the C20/EPA concentration is expanded upon here in order to present as complete a picture as possible of the capabilities (and limitations) of the technology in the fish oils application. Supercritical fluids do exhibit some very attractive separation properties, but they are not "magic solvents"; they cannot produce 90% EPA starting with the fish oil esters used by Eisenbach in the hot finger extractor he described.

Some simple material balance considerations permit the maximum achievable EPA concentration to be calculated using the data shown in Table III which gives the composition of the codfish oil esters that were tested by Eisenbach. As is calculated from the values given in the table, the C20 fraction comprises 26.7% (w/w) of the feed stock,, and the C20:5 (EPA) component comprises 54.7% of the C20's. Since to a first approximation supercritical extraction separates by carbon number, i.e., by the length or molecular weight of the fatty acid ester, the data in Table III show that the theoretical maximum EPA concentration achievable is 54.7%; if the C20's could be separated into a fraction containing only C20's, the EPA concentration in that fraction would be 54.7%. Eisenbach obtained a C20 fraction of 96.3% in a two pass system which is an excellent value; it is calculated that the EPA concentration in that fraction is 52.7%, also an excellent value relative to the theoretical value of 54.7%.

Several other studies on fish oils were reported (26-29) shortly after the Eisenbach paper attesting to the interest in supercritical fluid fractionation to concentrate EPA. The most

Table III. FATTY ACID PROFILES OF CODFISH OIL ESTERS

Component	Concentration (Wt%)
14:0	5.80
15:0	0.19
16:0	12.88
16:1	9.79
18:0	2.66
18:1	23.25
18:2	0.16
18:4	2.19
20:1	11.43
20:4	0.50
20:5	14.46
22:1	8.64
22:4	0.43
22:5	0.49
22:6	5.74

recent study of fish oils is the work of Nilsson et al., (38) at the National Marine Fisheries Service, Seattle, WA; their work has shown that, with a feed of urea fractionated menhaden oil ethyl esters, supercritical fluid fractionation can concentrate EPA to 96%. The concept and principles of urea fractionation and its ability to sequester and crystallize saturated and mono-ene esters from a mixture of esters have been covered elsewhere (39). As an example of the ability of urea fractionation to remove saturated and mono-ene components from a mixture of esters, Table IV lists the composition of whole menhaden oil esters and the esters that are produced by urea fractionation of whole esters. An examination of the compositions of the two ester feed stocks shows that the saturated and mono-unsaturated components have been almost quantitatively removed by urea fractionation; important for supercritical fluid fractionation considerations, the concentration of EPA theoretically achievable in the C20 fraction has been increased to 96.8%.

Figure 2 shows an extraction profile of EPA obtained during the batch continuous extraction of a charge of menhaden oil ethyl esters. Extraction conditions were 2200 psig, 40°C, with a 100°C heated zone (the "hot finger") located above the extractor, which as explained earlier, provides the reflux flow to enhance the separation of components. Each fraction collected was analyzed by capillary column gas chromatography for individual components. Each fraction that was collected represented less than 5% of the charge in order to simulate a continuous concentration profile of the components in the extract. As the figure shows, the profile has reached a maximum of 96% EPA concentration. The integrated average concentration of EPA in the fraction collected between points P_1 and P_2 is 90%, and the amount of EPA in this high concentration fraction represents a recovery, or yield, of 75% of the EPA that was initially present in the charge. (For emphasis again the concentration of EPA, not C20's, is 90%, and this high concentration is made possible by the use of the urea fractionated esters. Relative to the theoretical maximum of 96.8% calculated from the information in Table IV the 90% EPA concentration value obtained by Nilsson points out the excellent fractionation capabilities of supercritical fluids. Eisenbach, of course, showed the same in his studies when he achieved a 96% concentration of C20's in a fraction.)

Because of the high EPA concentration achieved in laboratory tests, supercritical fluid extraction is a potentially attractive process from scale-up considerations. A flow diagram of a continuous countercurrent extraction process that can produce a high concentration EPA product and a high concentration DHA (docosahexaenoic acid) product at very high yields is shown in Figure 3. In brief explanation of the process, a feed stream of ethyl esters is fed to the first extraction column of a two-column system where a separation between components lighter than the C20's and the combined C20/C22 components is made. The C20/C22 stream leaving the bottom of Column I is pumped to another column (Column II) where the C20's are separated from the C22's. Depending upon operating conditions (e.g., temperature, pressure, solvent to feed ratio, reflux of condensed esters

TABLE IV. FATTY ACID PROFILES OF MENHADEN OIL ETHYL ESTERS

	Composition, Weight %	
Component	Whole Esters	Urea Fractionated Esters
14:0	7.8	-*
16:0	15.6	-
16:1ω7	10.9	0.1
16:3ω4	1.1	5.3
16:4ω1	1.5	5.8
18:0	3.1	-
18:1ω9	7.5	-
18:1ω7	3.1	0.1
18:2ω6	1.3	0.2
18:3ω3	1.6	0.3
18:4ω3	3.0	7.6
20:1ω9	1.2	-
20:4ω6	1.0	1.4
20:4ω3	0.2	0.2
20:5ω3	16.5	48.6
21:5ω3	0.7	1.3
22:5ω3	2.5	0.9
22:6ω3	10.9	22.2
By carbon number**		
C14	8.1	-
C16	32.1	0.2
C18	21.3	14.0
C20	21.8	50.3
C22	14.2	27.4

* Below detectable level.
** Minor unidentified peaks have been assigned a carbon number by relative column retention time.

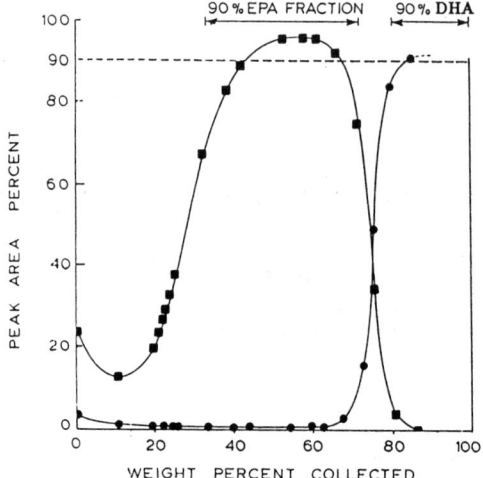

Figure 2. Extraction profile of eicosapentanoic acid from batch continuous extraction of menhaden oil ethyl esters. Extraction conditions: 2200 psig, 40 °C, with a 100 °C heated zone above extractor. (Reproduced with permission from Ref. 38. Copyright 1988 American Oil Chemists' Society.)

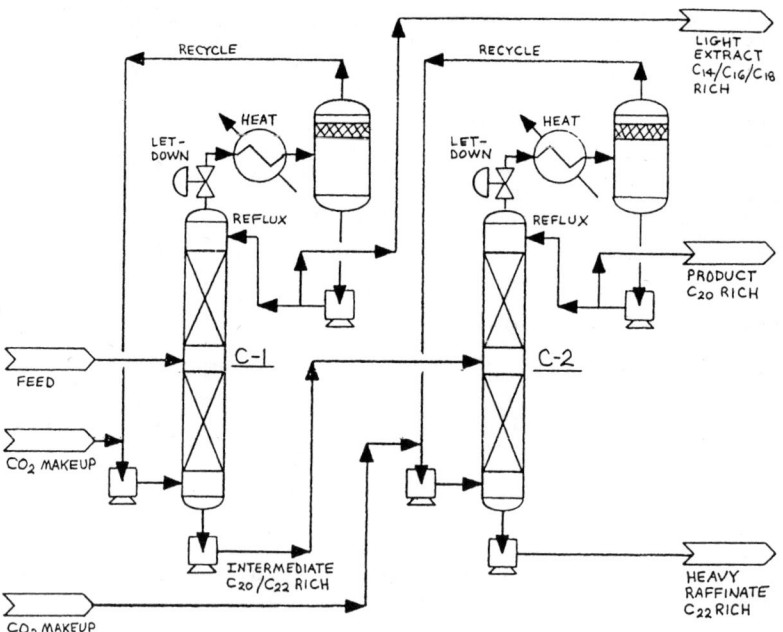

Figure 3. Flow diagram of a supercritical fluid extraction process for fish oil fractionation. (Reproduced with permission from Glitsch, Inc.)

returned to each column, and other factors) concentrations and yields of EPA and DHA can be made as high as desired in the respective streams. A detailed description of calculated plant performance based upon distribution coefficients and selectivities measured at the National Marine Fisheries has been reported elsewhere (40, 41).

Enzyme Catalyzed Reactions

Many biotechnology related extractions using supercritical fluids have been suggested and investigated, e.g., extraction of fermentation products such as acetic acid (42) and ethanol (43). Many of these were investigated experimentally during the period of emphasis of energy reduction, but as the cost of energy declined from its peak in the early 1980's, supercritical fluid extraction for its energy efficiency was found not ecenomically viable. (Interestingly, Francis had also published data on acetic acid- and ethanol-water-carbon dioxide ternary behavior, which could have been used to calculate distribution coefficients.) Other research studies of supercritical fluids in biotechnology areas are still on-going, and enzyme catalyzed reactions in supercritical fluids represent a subset within the broad range of biotechnology.

The first study of enzyme catalyzed reactions in a supercritical fluid was reported by Randolph et al. (44) who studied the hydrolysis of disodium p-nitrophenylphosphate hexahydrate using alkaline phosphatase. The workers carried out the enzymatic hydrolysis in carbon dioxide at conditions of 100 atm, $35^{\circ}C$ in an autoclave reactor. The disodium salt, enzyme, and a small amount of water were charged to the autoclave, and the extent of conversion determined as a function of elapsed reaction time. The authors explained their data by invoking that the rate determining step in the reaction is the dissolution of the disodium salt in supercritical carbon dioxide. (Other workers have, however, shown that the salt does not dissolve in carbon dioxide and that some other reaction path must be operative (45).)

A later paper reports on another enzyme catalyzed reaction, the oxidation of p-substituted phenols using polyphenol oxidase (46). Polyphenol oxidase is found in high concentration in mushrooms, some fruits, and in tea and tobacco leaves; it is the cause of the browning seen in bruised or broken plant tissues. The enzyme functions in plants to synthesize polyphenolic and quinoid compounds which through their antibiotic action inhibit microbiological contamination when the plants are damaged. These workers used a flow reactor; the configuration of the reaction vessel is shown in Figure 4. The reactant, p-cresol, was charged to a zone near the reactor inlet, and the enzyme was situated near the reactor outlet, the two zones separated by a barrier of glass wool. A continuous flow of carbon dioxide at 300 atm, $40^{\circ}C$, containing 2% oxygen and saturated with water (0.1%) was admitted to the reactor. The p-cresol was dissolved by the carbon dioxide and transported to the enzyme section where the reaction between the enzyme and the p-cresol dissolved in carbon dioxide took place; the flow reactor was used specifically to remove any ambiguity in the method of contact between the p-cresol and the enzyme. The stream leaving the enzyme section was expanded to ambient pressure causing the dissolved components to precipitate; they were collected in a trap situated in a dry ice-acetone bath.

The reaction sequence is shown in Figure 5. The dissolved p-cresol is enzymatically oxidized to the catecholic compound and in turn to the ortho quinoid species which usually exhibit intense red color. An intense, deep red solid was seen to precipitate within a few seconds after the start of the test pointing out that the reaction was quite rapid and also that the quinoid reaction products were soluble in carbon dioxide. The material that was collected in the trap (and the deep red material that was also found on the enzyme in the reactor at the end of the test) was analyzed by gel permeation chromatography; the chromatogram is shown in Figure 6. The molecular weight of the quinoid oligomers (formed because ortho and para quinoid compounds have a strong tendency to polymerize) is seen to range from about 8000 to about 100, the molecular weight of the "monomer", p-methyl orthoquinone, that was produced via enzyme catalysis.)

Another recent enzyme catalysis study investigated transesterification reactions with lipases (47). The reaction sequence involved loading triolein, free stearic acid, lipase, and a buffer solution of TES (N-tris(hydroxymethyl) methyl-2-aminoethane sulfonic acid) into a magnetically stirred autoclave, admitting carbon dioxide to a pressure of 137 bar and temperature of 35°C, stirring the contents for varying periods of time, at the end of which the contents were cooled with dry ice, the autoclave opened, and a sample taken. The sample was resolved into triglycerides and free fatty acids by thin layer chromatography, the spots scraped from the plate and converted to methyl esters, and analyzed by gas chromatography. A maximum of 16% replacement of oleic acid by stearic acid was measured. (In other tests it was found that mere exposure of the enzyme to supercritical carbon dioxide, i.e., without any reaction sequence, did not materially influence the activity of the lipase enzyme.)

Although not specifically fitting in the category of enzyme catalyzed reactions in supercritical fluids, there have been other studies that have investigated the effects of supercritical fluids on enzyme stability. In some cases there was a loss of activity when the enzyme was exposed to supercritical fluids, and the loss of activity was a desirable feature. In other studies, there was no loss of activity, and the retention of activity was advantageous.

As an example of the case where deactivation is advantageous, moist carbon dioxide at conditions of 600 atm, 80°C was found to improve the flavor of soy bean protein and to deactivate peroxidase enzymes that can result in subsequent undesirable oxidation of residual lipids in the protein meal (48); the deactivation of the peroxidase was, thus, beneficial since a better soy bean protein meal product resulted.

In another instance, the removal of lipids from mustard seed using supercritical carbon dioxide (49), the activity of the

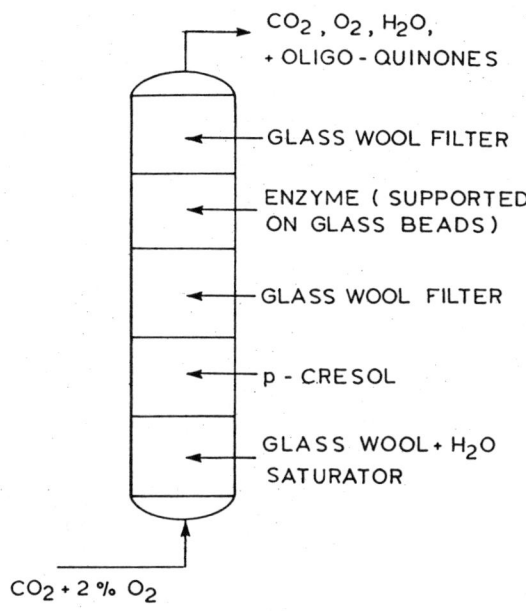

Figure 4. Flow reactor for the oxidation of p-substituted phenols using polyphenol oxidase.

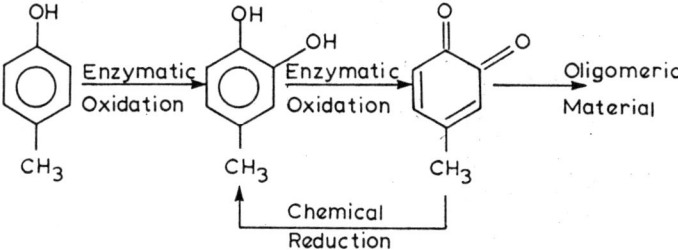

Figure 5. Reaction sequence of p-cresol to ortho quinoid catalyzed by polyphenol oxidase. (Reproduced with permission from Ref. 21. Copyright 1985 Humana Press, Inc.)

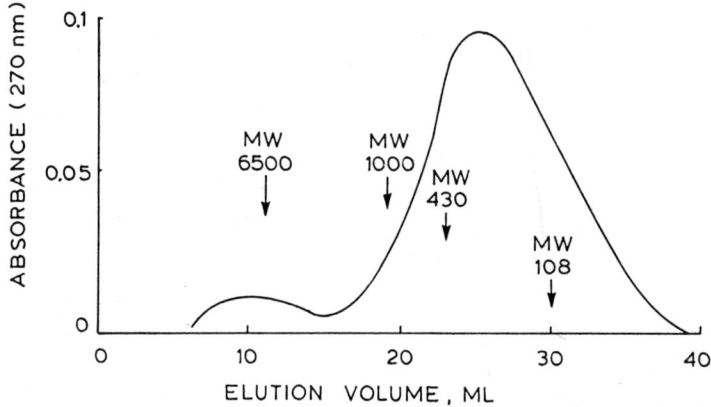

Figure 6. Elution pattern of polymeric material formed during oxidation of p-cresol by polyphenol oxidase in supercritical gas and hexane. (Reproduced with permission from Ref. 21. Copyright 1985 Humana Press, Inc.)

enzyme, myrosinase, in the mustard seed was retained and retention was a desirable result. Myrosinase is essential in forming one of the sensory compounds in mustard spice preparations. In the preparation of the spice, the oils must be removed from mustard seed to preclude separation and oxidation of oils in the final product. If hexane is used as the extraction solvent, it satisfactorily extracts all the oils; however, during the residual solvent removal step, the high temperature can deactivate the myrosinase resulting in a less satisfactory spice preparation. Supercritical carbon dioxide was found to be an excellent solvent, removing the undesired oils and simultaneously retaining all the activity of myrosinase.

Conclusion

As was brought out in this chapter, supercritical fluid extraction has been promoted perhaps somewhat over-zealously in the past, especially in its potential application to energy reduction. Although it can "do" many types of separations in energy related applications and elsewhere, it probably will not be economicaly viable in all cases. Many factors, not just technical feasibility, combine to influence the potential applicability of any process, and all factors must be evaluated; supercritical fluid extraction cannot be excepted from such careful analysis.
   A number of applications for supercritical fluid extraction have progressed to commercial scale, many are in advanced pilot plant or test market evaluation, and as the capabilities of the technology are becoming better understood and the technology emerging out from under the cloud of initial hype and misperception, new applications are increasingly being investigated as shown in Table II. The two applications that were described in this chapter were selected somewhat arbitrarily, and they have not yet been subjected to the careful evaluation suggested here. The supercritical fluid fractionation of fish oil esters works very well from technical considerations, and it can be competitive economically if the marketplace demands a 90+ % EPA product. As another example, de-oiling mustard seed by supercritical carbon dioxide which retains the activity of the enzyme, myrosinase, may become viable if greater performance demands are placed on the spice preparation. Case-by-case evaluations are always required; by such evaluation viable applications for supercritical fluid extraction are increasingly being identified.

Acknowledgments

I gratefully acknowledge the efforts of Dr. Bonnie Charpentier in the preparation of this manuscript as published here and those of the ACS Books Department in the preparation of the figures.

Literature Cited

1. Krukonis, V.J. 188th ACS Mtg. Philadelphia, August 1984.

2. Hannay, J.B., Hogarth, J. *Proc. Roy. Soc. (London)* 1879, 29, 324.
3. Diepen, G.A.M., Scheffer, F.E.C. *J. Am. Chem. Soc.* 1948, 70, 4081.
4. McHugh, M.A., Mallet, M.W., Kohn, J.F. *In Chemical Engineering at Supercritical Fluid Conditions* Ann Arbor Science, Ann Arbor, MI, 1983; 113.
5. Paulaitis, M.E., Kander, R.G., DiAndreth, J.R. *Ber. Bunsenges. Phys. Chem.* 1948, 88, 869.
6. Krukonis, V.J. *Nat. AIChE Mtg.* Boston, August 1979.
7. Brunner, G., Peter, S. *Ger. Chem. Eng. (Engl. Transl.)* 1982, 5, 181.
8. Friedrich, J., List, G., Heakin, A. *J. Am. Oil Chem. Soc.* 1982, 59, 288.
9. McHugh, M.A., Krukonis, V.J. *Supercritical Fluid Extraction: Principles and Practice* Butterworth Publishing, Boston, MA, 1986, Ch. 2.
10. Francis, A.W. *J. Phys. Chem* 1954, 58, 1099.
12. Krukonis, V.J. *Ann. AIChE Mtg.* San Francisco, November 1984.
13. Larson, K.A., King, M.L. *Biotech. Prog.* 1986, 2, 73.
14. Krukonis, V.J. *Nat. AIChE Mtg.* Dallas, March 1983.
15. Kilgor, I., McGrath, J.E., Krukonis, V.J. *Polym. Bull.* 1984, 12, 491.
16. Krukonis, V.J. *Polymer News* 1985, 11, 7.
17. Kumar, S.K., Suter, U.W., Reid, R.C. *ACS Fuel Chemistry Preprints* 1985, 30, 66.
18. Modell, M., deFilippi, R.D., Krukonis, V.J. *Ann. ACS Mtg.* Miami, November 1978.
19. Modell, M. 1982, U.S. 4,338,199.
20. Randolph, T.W., Blanch, H.W., Prausnitz, J.M., Wilke, C.R. *Biotech Letters* 1985, 7, 325.
21. Hammond, D.A., Karel, M., Klibanov, A.M., Krukonis, V.J. *Appl. Biochem. Biotech.* 1985, 11, 393.
22. Taneguchi, M., Kamihira, M., Kobayashi, T. *Agric. Biol. Chem.* 1987, 51, 593.
23. Coenan, H., Hagen, R., Knuth, M. 1983, U.S. 4,400,398.
24. Shutz, E., Vollbrecht, H.R., Sandner, K., Sand, T., Muhlnickel, P. 1984, U.S. 4,470,927.
25. Eisenbach, W. *Ber. Bunsenges. Phys. Chem.* 1984, 88, 882.
26. Krukonis, V.J. *75th Am. Oil Chem. Soc. Mtg.* Dallas, May 1984.
27. Suzuki, Y., Shimazu, M., Arai, K., Saito, S. *Ann. AIChE Mtg.* Miami Beach, November 1986.
28. Eisenbach, W. *190th ACS Mtg.* Chicago, September 1985.
29. Nilsson, W.B., Stout, V.F., Hudson, J.K., Spinelli, J., Gauglitz, E.J., Jr. *77th AOCS Mtg.* Honolulu, May 1986.
30. Anon. *Newsweek* 1986, July 14, p. 5.
31. Zosel, K. 1974, U.S. 3,806,619.
32. Laws, D.R.J., Bath, N.A., Ennis, C.S., Wheldon, A.G. 1980, U.S. 4,218,491.
33. Cookson, C.B. *Spring Tech. Sym. Bulk Pharma. Chem.* Newport, May 1987.

34. Perkins, B. The Cincinnati Enquirer January 10, 1987.
35. Anon. Ind. Chem. News 1986, September, 11.
36. Rizvi, S.S.H., Benado, A.L., Zollweg, J.A., Daniels, J.A. Food Tech. 1986, June, 55.
37. Rizvi, S.S.H., Daniels, J.A. 193rd ACS Mtg. Denver, April 1987.
38. Nilsson, W.B., Gauglitz, E.J., Jr., Hudson, J.K., Stout, V.F., Spinelli, J. J. Am. Oil Chem. Soc. 1988, 65, 109.
39. Sumerwell, W.N. J. Amer. Chem. Soc. 1957, 79, 3411.
40. Krukonis, V.J. Design of a Continuous Countercurrent Extraction Process for Producing Concentrated EPA Ethyl Esters 1987, Appendix submitted to National Marine Fisheries Service, Seattle, WA.
41. Krukonis, V.J. 194th ACS Mtg. New Orleans, September 1987.
42. Shimshick, E.J. 1981, U.S. 4,250,331.
43. Moses, J.M., Goklen, K.E., deFilippi, R.D. Ann. AIChE Mtg. Los Angeles, November 1983.
44. Randolph, T.W., Blanch, H.W., Prausnitz, J.M., Wilke, C.R. Biotech. Letters 1985, 7, 325.
45. Krukonis, V.J., Hammond, D.A. Submitted to Biotech. Letters 1987.
46. Hammond, D.A., Karel, M., Klibanov, A.M., Krukonis, V.J. Appl. Biochem. Biotech. 1985, 11, 393.
47. Nakamura, K., Chi, Y.M., Yamada, Y., Yano, T. Chem. Eng. Commun. 1986, 45, 207.
48. Christianson, D.D., Friedrich, J.P. 1985, U.S. 4,495,207.
49. Taneguchi, M., Nomura, R., Kijima, I., Kobayashi, T. Agric. Biol. Chem. 1987, 51, 413.

RECEIVED January 29, 1988

Chapter 3

Analytical Supercritical Fluid Extraction Methodologies

Bob W. Wright, John L. Fulton, Andrew J. Kopriva, and Richard D. Smith

Chemical Sciences Department, Pacific Northwest Laboratory, Richland, WA 99352

> Off-line supercritical fluid extraction, ultrasonic supercritical fluid extraction, and on-line supercritical fluid extraction-gas chromatography methodologies that have been developed specifically for analytical sample preparation and analysis are described. These methods offer the potential for extraction rate increases of over an order of magnitude, and are compatible with on-line analysis which provides the basis for automated sample preparation and analysis. These methods are particularly useful for small sample sizes or trace levels of analytes, and have been demonstrated to operate quantitatively. Combined ultrasonic supercritical fluid extraction can further enhance extraction rates from macro-porous materials by inducing convection through internal pores. The apparatus and instrumentation are described in detail and several examples are presented illustrating the applicability of these methodologies.

Until recently, the use of supercritical fluids for sample extraction was generally confined to chemical processing applications (1-3). However, the use of supercritical fluid extraction (SFE) for analytical purposes is attracting increased attention and the development of several new techniques has been reported (4-14). A number of potential advantages are possible with SFE compared with conventional extraction methods. These advantages include more rapid extraction rates, the possibility of more efficient extractions, increased selectivity, possible analyte fractionation during extraction, and compatibility with on-line analysis methods such as continuous spectroscopic monitoring or periodic chromatographic analyses.

The potential advantages of SFE accrue from the properties of a solvent at temperatures and pressures above its critical point. At elevated pressure this single phase will have properties that are intermediate between those of the gas and liquid phases and are dependent on the fluid composition, pressure, and temperature. The compressibility of supercritical fluids is large just above the critical temperature and small changes in pressure result in large changes in density of the fluid (3). The density of a supercritical fluid is typically 100 to 1000 times greater than that of the gas and solvating characteristics approaching those of a liquid are imparted. However, the diffusion coefficients and viscosity of the fluid remain intermediate between those of the gas and liquid phases at moderate densities (3), thus allowing rapid mass transfer of solutes compared with the liquid. Many fluids also have comparatively low critical temperatures that allow extractions to be conducted at relatively mild temperatures, e.g., 31 °C for carbon dioxide. In addition to controlling the fluid pressure and/or temperature to regulate the density or solvating power, one can use various fluids or fluid mixtures that exhibit different specific chemical interactions to obtain the desired solvent strength and selectivity. The liquid-like solvent power and rapid mass-transfer properties of supercritical fluids clearly provide the potential for more rapid extraction rates and more efficient extraction than is feasible with liquids due to better penetration of the matrix.

This paper describes various analytical supercritical fluid extraction methodologies developed in the authors' laboratory and summarizes selected investigations conducted to evaluate the applicability and efficiency of these methods for a range of applications. Described methodologies include off-line supercritical fluid extraction, ultrasonic supercritical fluid extraction, and on-line supercritical fluid extraction-gas chromatography.

Off-line Supercritical Fluid Extraction

Sample preparation is often more difficult and time-consuming than the actual analysis procedure. Furthermore, extraction of analytes from the matrix is generally the most time-consuming step of sample preparation and it can lead to relatively inefficient analyte recoveries. Off-line supercritical fluid extraction provides an alternative to traditional Soxhlet or ultrasonic liquid extraction methods. Several recent studies have shown analytical SFE provides comparable or better extraction efficiencies than Soxhlet extraction (8,12,14). More importantly, increased extraction rates of over an order of magnitude were achieved (12), which offer significant time savings. Off-line SFE is adaptable to sample sizes ranging from a few milligrams to several grams and is equally applicable to the recovery of trace analytes or complex mixtures. Examples of both types of applications are described.

Apparatus. The off-line SFE instrumentation consisted of three main components: a high-pressure pump, a heated extraction cell, and a depressurization and sample collection system. A schematic diagram of this instrumentation is shown in Figure 1. A modified Rabbit HPX solvent delivery system equipped with a 10 mL/min pump head and an electronic pressure monitor (Rainin Instrument Co., Woburn, MA) was used to pressurize and deliver the extracting fluids. The pump head, check valves, and several inches of the inlet line immediately prior to the pump head were cooled by recirculating a -15 °C ethylene glycol-water mixture through copper blocks that were machined to fit the pump head geometry. The inlet check valve cartridge was also modified for direct connection to 1/8-in. stainless steel tubing to allow a larger supply of low-pressure solvent to enter the pump. No other modifications to the check valves or piston assembly were required. High-pressure syringe pumps with adequate volumes have also been used (7,14).

The pressurized fluid was transferred to the 316 stainless steel high-pressure extraction vessel (shown in Figure 2) through 1/16-in.

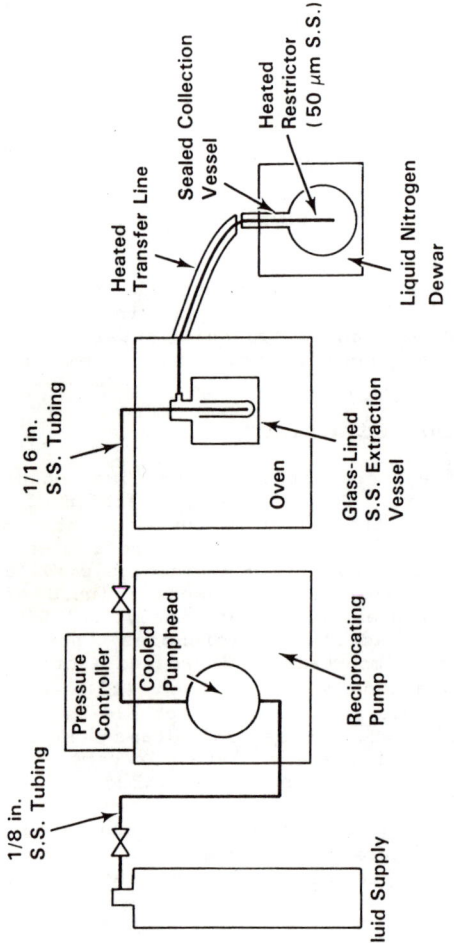

Figure 1. Schematic diagram of the off-line supercritical fluid extraction apparatus. (Reproduced from Ref. 12. Copyright 1987 American Chemical Society.)

3. WRIGHT ET AL. *Supercritical Fluid Extraction Methodologies* 47

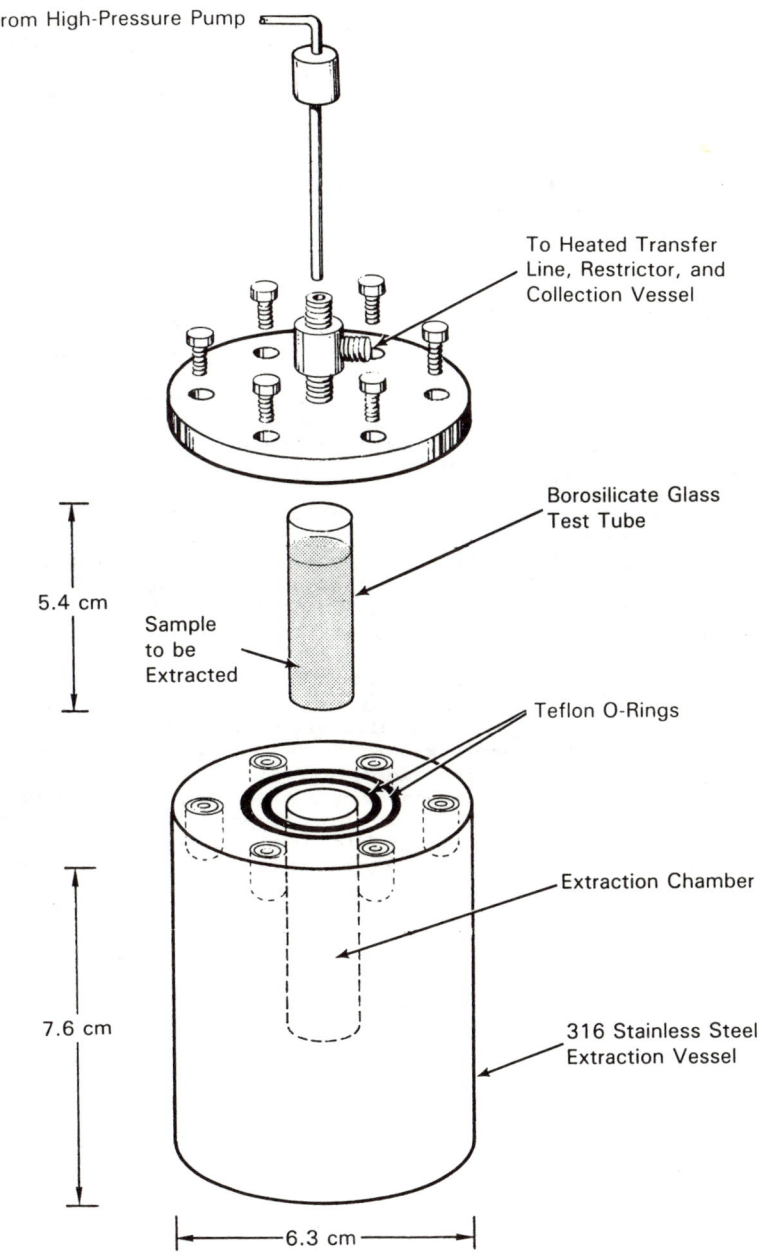

Figure 2. Design of supercritical fluid extraction vessel. (Reproduced from Ref. 12. Copyright 1987 American Chemical Society.)

stainless steel tubing. The extraction cell was constructed to operate safely above 400 bar and 200 °C. The extraction cell body and top were sealed with double Teflon O-rings. The transfer line from the pump extended to the bottom of the extraction cell chamber (~5 mL volume) where the sample was placed in a 5.4 x 1.3 cm o.d. borosilicate glass test tube. This design allowed the extraction fluid to sweep the sample from bottom to top and then exit the extraction vessel. A stainless steel frit with 2.0-μm pores was placed in the exit port of the extraction vessel to prevent sample from being flushed out in case of sudden pressure surges. Empty 1/4-in. o.d. HPLC columns of the appropriate length to give the desired volume have also be used as extraction cells. The extraction vessel was maintained at elevated temperatures in a Hewlett-Packard 5700 gas chromatograph oven. The oven was continuously purged with nitrogen when using flammable extraction fluids or fluid mixtures. The extract stream was transported from the extraction cell to the exterior of the oven through 1/16-in. i.d. glass-lined stainless tubing. The transfer line was connected to a 10-12-cm length of 50-μm i.d. stainless steel tubing which was crimped at the exit end to control the flow rate of the fluid and serve as a depressurization zone for the extraction stream. Short lengths (5-10-cm) of 10-25-μm i.d. fused silica tubing have also been used as restrictors.

The extraction effluents were collected by freezing (or liquefying) in a sealed round bottom flask cooled in a liquid nitrogen bath. This method ensured that no analyte losses to the atmosphere occurred during collection. Collection can also be done by bubbling the extraction effluent into a vial containing a few mL of a solvent or internal standard solution (7,14).

Trace Level Extraction. To provide a challenging sample for evaluation of off-line SFE, activated carbon was spiked at 50 ppm with several polar and higher molecular weight polycyclic aromatic compounds. A one-gram sample was subjected to 16 h of Soxhlet extraction using carbon disulfide and then followed with a second similar extraction using methylene chloride. Another sample was extracted for 1 h with supercritical carbon dioxide at 125 °C and 400 bar. As shown in Table I, no detectable levels of the compounds were recovered in the combined Soxhlet extracts. However, low levels of the compounds were recovered with supercritical carbon dioxide extraction of the activated carbon. Although only low levels of the

Table I. Extraction Comparison of Activated Carbon by Soxhlet and SFE

Compound	Soxhlet % Recovery	SFE % Recovery
Chrysene	0	1
Benzanthrone	0	6
1-Nitropyrene	0	10
Dibenzocarbazole	0	10
Coronene	0	0.2

spiked analytes were recovered from the carbon, this example illustrates the potential of SFE for extracting low-level analytes from highly adsorptive matrices.

Important considerations in analytical SFE are the provisions taken for collecting the sample during the depressurization process. Depending on the exact conditions, it is possible for the analytes to nucleate and become entrained in the expanding gas and form an aerosol which can be easily lost to the atmosphere. Experiments have shown that over 90% of trace level analytes can be lost under some conditions when using open collection in a narrow-necked flask cooled to subambient temperatures (12). Collection in a sealed vessel eliminates losses due to either volatility or to aerosol formation, but increases the experimental complexity. Collection by bubbling the extraction effluent in a few milliliters of solvent is also effective for low fluid flow rates. The cooling due to expansion of the fluid minimizes volatility losses of both the analyte and solvent. However, this collection method becomes impractical for the higher fluid flow rates needed for rapid extraction of larger sample sizes (e.g., gram range), since very high gas flow rates would be required making it difficult to contain the flow in a small volume of solvent. The volume of a fluid increases by approximately three orders of magnitude during expansion to a gas. Experience has shown that at least ten sample volumes (or extraction vessel volumes) of compressed fluid are generally needed to achieve "exhaustive" extraction for typical matrices. Obviously, analytes with very low solubility in supercritical fluids could require larger extraction volumes.

Complex Mixture Extraction. Analytical SFE can also be used for complex mixture sample preparation. Typical examples using hazardous waste samples are described below. Sample A was a soil boring contaminated with coal gasification residuals and sample B was from a waste stream from a treatment facility. The major objective of these studies was to compare the extraction abilities (e.g., amount of material extracted) of three different fluid systems using approximately four-gram aliquots of the samples. The specific fluid systems, the extraction conditions, and the percentage of the total mass of material extracted from each sample are listed in Table II.

Table II. Supercritical Fluid Extraction Comparison of Hazardous Waste Samples

Fluid System	Extraction Conditions			Percent Extracted	
	Temp. (°C)	Pressure (Bar)	Fluid Volume (mL)	Sample A	Sample B
Carbon Dioxide	150	415	500	27	28
Carbon Dioxide-Methanol (80:20)	150	400	500	26	49
Pentane-Ethanol (93:7 mol %)	215	165	750	29	51

The percentage of material extracted with the three different fluid systems was very similar for sample A. However, the amount of material removed from sample B was approximately 1.8 times greater with methanol-modified carbon dioxide and ethanol-modified pentane than with pure carbon dioxide. The higher extraction temperature used for the pentane system could have contributed to the improved efficiency for this fluid. Identical temperatures were used for the carbon dioxide and carbon dioxide-methanol fluid systems, which suggests that the components were more soluble in the more polar fluid system. This is consistent with the sample composition, since subsequent analyses of sample B indicated that it contained phenolics, amines, esters, nitrated aromatics and other polar compounds. Sample A, on the other hand, contained neutral polycyclic aromatic compounds that would not be expected to have significantly higher solubility in the more polar fluid systems. However, the gas chromatographic profiles of the carbon dioxide extract and the carbon dioxide-methanol extract of sample A were slightly different. As can be observed from the chromatograms in Figure 3, there are greater relative quantities of the higher molecular weight compounds in the carbon dioxide-methanol extract than the pure carbon dioxide extract. This may be accounted for by slightly greater solubility of these higher molecular weight compounds in the more polar fluid. These examples illustrate the potential of analytical SFE for rapid and efficient complex mixture sample preparation.

Ultrasonic Supercritical Fluid Extraction

The application of ultrasound during supercritical fluid extraction creates intense sinusoidal variations in density and pressure which have the potential of improving solute diffusion and enhancing overall extraction rates. For a given power input, it can be predicted that the amplitude of the resulting density waves will be approximately twice as great as in a gas or liquid, whereas the pressure amplitude will be intermediate between that of a gas and liquid. These density waves would be expected to induce convection of solutes from the inner pores of a material since the density waves in the matrix material will be much smaller than that of the fluid in the pores. Furthermore, acoustic streaming, the characteristic pattern of steady micro-vortices in sonicated liquids (15), also occurs in supercritical fluids and would likewise be expected to decrease external mass transfer resistance, the convective transport of the solute from the surface of the matrix into the bulk solution, and to create convection in the macropores of the material by recirculation of the fluid. The application of ultrasound to supercritical fluids also creates localized heating that could enhance the desorption rate of solutes from a matrix. Sound energy is absorbed in a supercritical fluid and near the critical point, attenuation of the sound waves increases dramatically because of higher viscous dissipation and structural relaxation of molecular clusters (16). It should be noted that cavitation (the creation and violent collapse of small vapor bubbles), the predominate effect of ultrasound in normal liquids, does not occur in supercritical fluids because no vapor-liquid surface exists to sustain the bubble structure. However, sonication may also improve extractions by altering or disrupting the internal structure of a porous material.

Apparatus. The ultrasonic supercritical fluid extraction vessel consisted of a 25 mL high-pressure stainless steel cell equipped with two sets of 0.75-in. diameter windows (for observation during extraction) and a commercial 20 kHz high power ultrasonic horn (Branson W-350 Sonifier with a 102 converter). The cell geometry was designed to maximize the sonic energy per unit volume and to create resonance effects to better transfer energy from the sonic horn to the fluid. The usable cell volume underneath the horn was approximately 12 mL. The ultrasonic transducer was connected to the extraction vessel through an O-ring seal at a wave node to minimize dampening of the sonic energy.

The extraction vessel was placed in a thermostated water bath, and the fluid inside the vessel was recirculated through a coil of 1/8-in. o.d. stainless steel tubing with a small magnetically

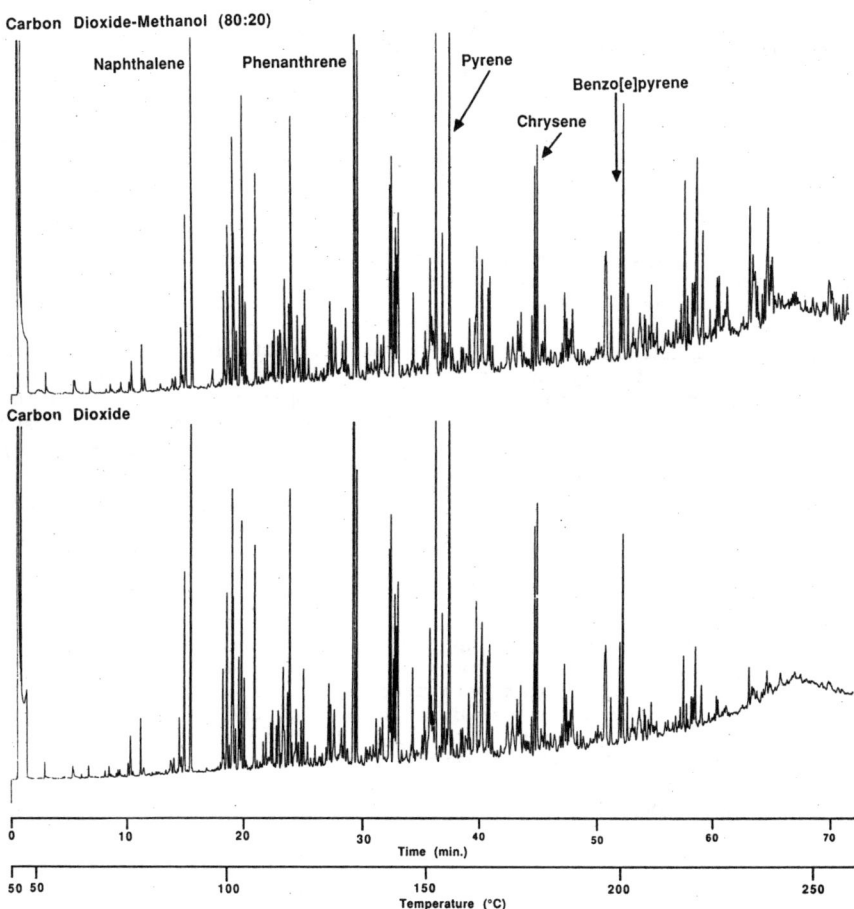

Figure 3. Capillary GC chromatograms of the supercritical carbon dioxide (bottom) and methanol modified carbon dioxide (top) extracts of a hazardous waste sample.

coupled gear pump (183-346, Micro-Pump Corp.). Fluid recirculation provided a convenient method of obtaining controlled mixing and also served to dissipate the heat generated from the sonic energy. A schematic diagram of the ultrasonic extraction system is shown in Figure 4. The concentration of the extracted analytes was directly monitored by metering a low flow of the extraction effluent through a 254 nm UV absorbance detector (Altex, Model 153) equipped with a high pressure flow cell. The extraction effluent flow rate was regulated with a fused silica capillary restrictor attached to the outlet of the detector. The flow rate through the detector was low enough so that, with continuous absorbance measurements, only a small percentage of the overall system fluid volume was lost. The extraction effluent was collected by inserting the capillary restrictor exit into a vial containing a suitable solvent. The extraction system was initially pressurized with a modified Varian 8500 syringe pump. By controlling the pump pressure, a small make-up stream of fluid was introduced to compensate for the loss of fluid through the UV detector.

<u>Ultrasonic SFE of Adsorbents</u>. To evaluate ultrasonic SFE, the extraction rates of chrysene from various porous adsorbent materials using supercritical carbon dioxide were compared before and during the application of ultrasound. Chrysene was chosen as test analyte because of its low solubility in carbon dioxide ($<50 \times 10^{-5}$ mol/L) and its high UV absorbance ($E_{254} \approx 6 \times 10^4$ L/mol-cm in a liquid) which minimizes the effect of solute concentration on the diffusion coefficient and allows low concentrations to be easily detected. Tenax, alumina, and activated charcoal were used as adsorbents to offer a range of pore sizes and adsorbent strengths. The chrysene molecular size would not be expected to impede diffusion out of the pores, and it would be expected to be more strongly adsorbed by the activated charcoal than by the alumina or Tenax. Relatively mild extraction conditions with temperatures ranging from 35 to 50 °C and pressures from 85 to 135 bar were used in this study.

Experiments were conducted by allowing the SFE to reach steady-state conditions and then applying ultrasound. Changes in extraction rate could be determined from changes in the slope of the UV absorbance as a function of time. At low recirculation pump flow rates (<100 mL/min), significant improvements in extraction rates were obtained with the application of ultrasound. Since the temperature remained essentially constant (within 2 or 3 degrees), these results indicated that the external mass transfer resistance, the diffusion of the analyte from the outside surface of the adsorbent into the bulk solution, significantly retarded the rate of extraction. It is not surprising that sonication would significantly decrease external mass transfer resistance since vigorous microstreaming was clearly evident from visual observations made during the extraction. In fact, ultrasound power levels were limited to 20 watts to prevent disruption of the adsorbent matrices. By recirculating the fluid at a high enough flow rate (>500 mL/min), the external mass transfer resistance was reduced to a negligible level (higher flow rates did not increase the extraction rate) and the effects of ultrasound on only the internal diffusion and desorption rates could be evaluated. The combined diffusion and desorption rates of chrysene from the three adsorbents are listed in Table III. As can be observed, no increases in the extraction rates were obtained under these conditions with the application of moderate ultrasonic power levels. It should also be noted that chrysene extraction rates were not being limited by solubility. It can be concluded that 20 kHz ultrasound at the power levels employed did not enhance the desorption or convection of chrysene from the micro-pores of the adsorbents. However, it appears that for these matrices that the application of ultrasound can be useful for enhancing extraction rates when conventional agitation or stirring methods are precluded (e.g., microextraction cells).

Table III. Chrysene Extraction Rate In Supercritical Carbon Dioxide

Adsorbent	Pore Size (nm)	Loading (mole/mL)	Extraction Rate (mole/min-mL)	Ultrasound Rate (20 w at 20 kHz)
Tenax	200	2×10^{-7}	1×10^{-9}	1×10^{-9}
Alumina	20	2×10^{-7}	1×10^{-9}	1×10^{-9}
Carbon	10	20×10^{-7}	0.1×10^{-9}	0.1×10^{-9}

<u>Ultrasonic SFE of Roasted Coffee Beans</u>. The extraction rate of caffeine from roasted coffee beans was also evaluated and provides an interesting contrast to the micro-porous adsorbents. The coffee beans had pores sizes ranging from 10 to 100 µm. The extraction rate with supercritical carbon dioxide at 40 °C and 103 bar pressure was monitored for 3 h, and then ultrasound was also applied for several additional hours. The profile from the UV detector of this extraction is shown in Figure 5. Without ultrasound the extraction rate increased for the first 30 min of the extraction and then leveled off to a constant value. After application of ultrasound the extraction rate increased by a factor of eight for a short time and then decreased to a lower, but significantly higher rate than prior to ultrasound application. Samples of the extraction effluent were also collected and analyzed by capillary gas chromatography. Caffeine was found to be the major component of the extracts and the quantities collected over equal time periods were proportional to the observed extraction rate changes. The fluid recirculation rate was held sufficiently high during the extraction to essentially eliminate the external mass transfer resistance. Consequently, the higher extraction rate obtained with ultrasound can probably be attributed to induced convection through the pores of the beans. The enhanced extraction rate may have also resulted from disruption of internal cell membranes. However, no noticeable degradation of the outer shells of the beans could be visually observed.

In summary, preliminary investigations suggest that the application of ultrasound during SFE provides an efficient mechanism of vigorously stirring and thus decreasing the external mass transfer resistance of the sample particles. This type of agitation may be particularly useful for small micro-scale extraction cells where more traditional means of stirring would be difficult or impossible. The application of ultrasound may also induce convection and enhance transport of solutes through the internal structure of macro-porous materials increasing the overall extraction rates of the analytes.

<u>On-line Supercritical Fluid Extraction-Gas Chromatography</u>

A logical extension of supercritical fluid extraction in chemical analysis is to combine the process with a chromatographic method. The variable solvating power of a supercritical fluid provides the mechanism for the selective extraction of the components of interest

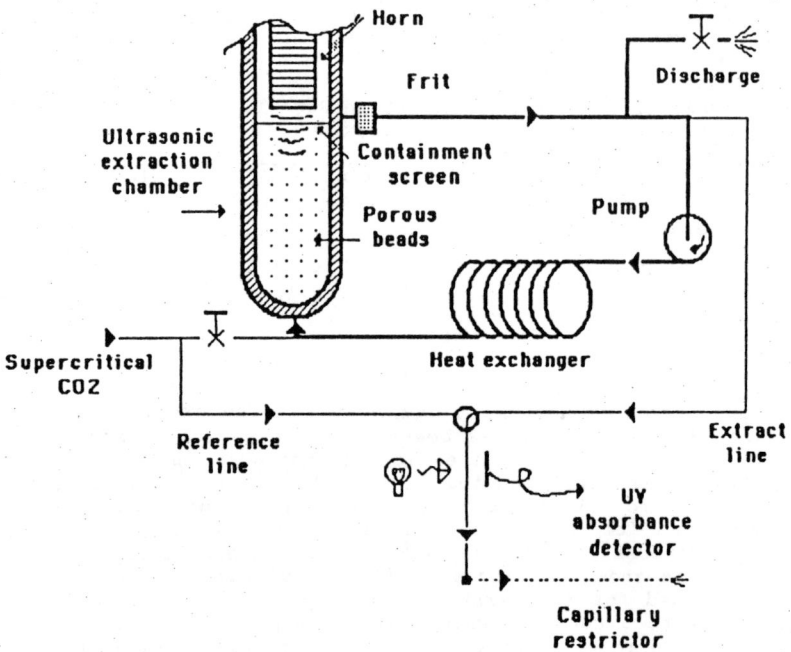

Figure 4. Schematic diagram of the ultrasonic supercritical fluid extraction apparatus.

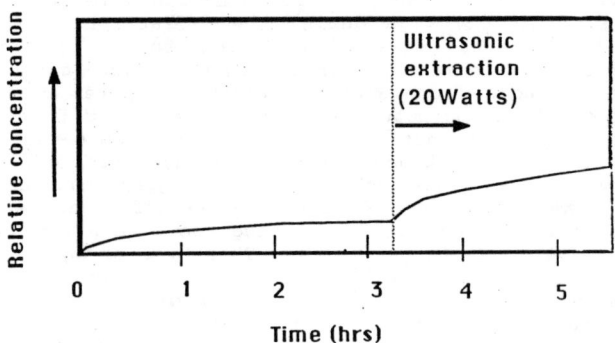

Figure 5. UV absorbance extraction profile of roasted coffee beans.

from the sample matrix and provides the basis for an automated method where sample preparation and analysis can be instrumentally linked. The on-line extraction-analysis approach is particularly attractive for small sample sizes and/or trace analysis where low levels of analytes are present. Various types of on-line analyses have been reported and include continuous monitoring of the total SFE effluent by mass spectrometry (9), combined SFE-high performance liquid chromatography (4), combined SFE-packed column supercritical fluid chromatography (6), and on-line SFE-gas chromatography (13).

Capillary gas chromatography was used for the analysis method in this study because of its relatively simple operation and its capability of providing high-resolution separations. In addition, the vast majority of compounds that are extractable with nonpolar fluids, such as carbon dioxide, are amenable to gas chromatographic analysis. Two general modes of operation are possible with the on-line SFE-gas chromatography approach: 1) quantitative extraction and analysis (where a matrix is exhaustively extracted and analyzed with a calibrated instrument) and 2) single or multiple selective extractions of a sample matrix to obtain qualitative (and potentially quantitative) analysis of specific fractions. Both of these modes of operation are described.

Apparatus. The automated on-line SFE-gas chromatography instrumentation consisted primarily of four sections which included a high-pressure pump and extraction cell, a switching valve and interface region, a gas chromatograph with a flame ionization detector, and a minicomputer and its associated interface circuitry. A schematic diagram of this instrumentation is shown in Figure 6.

A modified Varian 8500 syringe pump provided a high pressure supply of carbon dioxide to the extraction cell. The carbon dioxide was purified by distilling through activated charcoal while filling the syringe pump. Microextraction cells (see Figure 7) were constructed from Swagelok stainless steel zero volume 1/4-in. to 1/16-in. column end fittings (SS-400-6-1ZV) containing two 1/4-in. o.d. sintered stainless steel frits with 2.0 μm mean pore size separated by a 1/8-in. long x 1/4-in. o.d. x 3/32-in. i.d. stainless steel insert. The 1/4-in. o.d. inlet to the extraction cell was made by cutting and silver soldering the smoothed end from standard 1/16-in. o.d. stainless steel tubing that was inserted through a short length (1-2-in.) of 1/4-in. o.d. x 5/64-in. i.d. stainless steel tubing. This design provided an entirely stainless steel extraction cell with a total volume of approximately 15 μl (excluding internal frit volumes). Larger cell volumes were obtained by using larger i.d. tubing and/or longer inserts.

The extraction cell and several inches of inlet tubing were placed inside a thermostatically regulated heating block to control the fluid and extraction cell temperature. Extraction temperatures of 40-50 °C were generally used. An air-actuated Rheodyne 7010 six-port switching valve was used to direct the extraction cell effluent to either an exterior collection reservoir or to the gas chromatographic column for on-column deposition and concentration of the extract. Short lengths (2-4-cm) of 5-25-μm i.d. fused silica were used to regulate the fluid flow rate and to allow depressurization of the fluid to a gas. The decompressed gas flow rate of the extraction fluid ranged from 50 mL/min to 300 mL/min (~50-300 μl/min fluid flow) and several extraction cell volumes of pressurized fluid were generally used for an extraction. The restrictor for the on-column deposition was mounted through a 1/16-in. tee to allow the gas chromatograph carrier gas to enter coaxially along the restrictor. This connection was also located in a heated block to

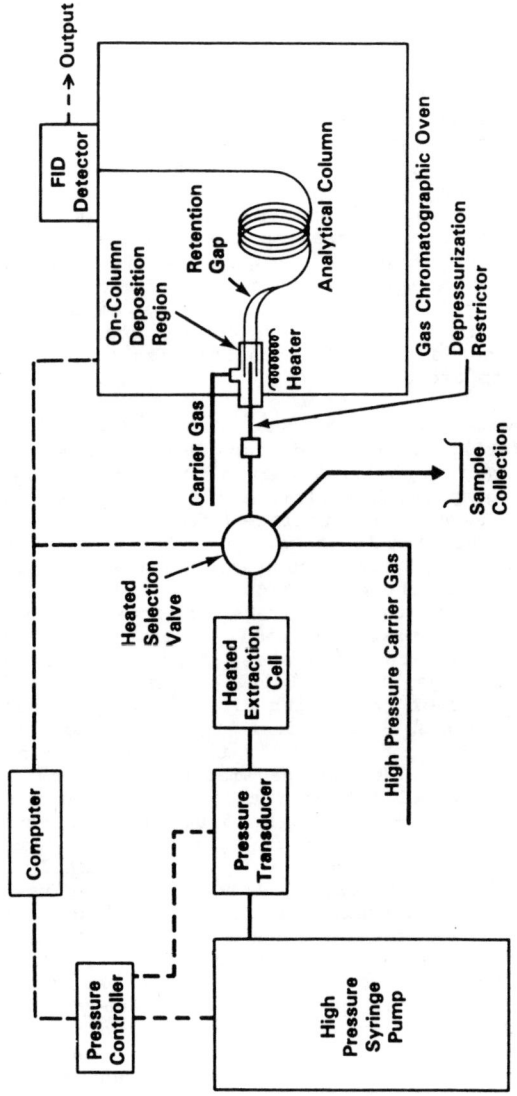

Figure 6. Schematic diagram of the on-line supercritical fluid extraction-gas chromatography instrumentation. (Reproduced from Ref. 13. Copyright 1987 American Chemical Society.)

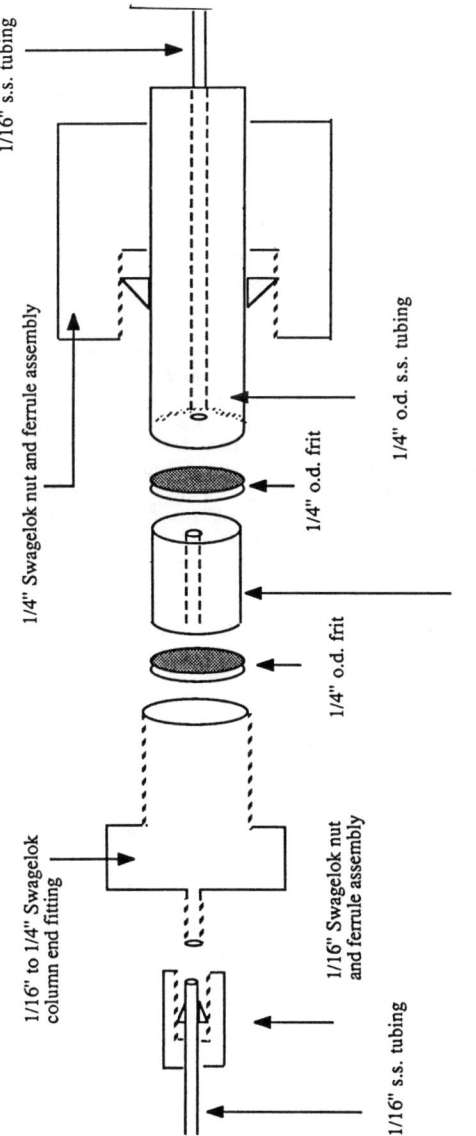

Figure 7. Design of the microextraction cells. (Reproduced from Ref. 13. Copyright 1987 American Chemical Society.)

control the temperature of the restrictor and expansion region. Typically, this region was maintained near the upper operating temperature of the chromatographic oven (between 250 and 280 °C).

A Hewlett-Packard 5890 gas chromatograph equipped with a flame ionization detector was used to perform chromatographic analyses. Fused silica chromatographic columns were either 15 m x 0.25 mm i.d. or 15 m x 0.53 mm i.d., coated with 0.25-μm or 5.0-μm film-thickness, respectively, of DB-5 stationary phase (J&W Scientific). A short retention gap of deactivated fused silica tubing (30 cm x 0.53 mm i.d.) was connected to the inlet of the chromatographic column to aid solute focusing and concentration of the extraction effluent. The oven was cooled to subambient temperatures using carbon dioxide during on-column deposition to aid focusing and concentration of the extraction effluent. Subsequent analyses were performed by temperature programming at selected rates. Helium was used for the carrier gas at linear velocities of approximately 40 cm/sec.

Qualitative Extraction Analyses. Qualitative characterization of a sample matrix can be obtained by periodic sampling and analysis of the extraction effluent at various pressures. Typically, a sample is extracted at progressively higher pressures to obtain selective fractions. However, to achieve maximum selectivity, the sample must be exhaustively extracted at each pressure to remove essentially all the material that is soluble at a given pressure prior to the next higher extraction pressure. Rigorous execution of this process can be difficult with the current instrumentation, so less selective fractionation is usually adopted for qualitative analyses.

An example of the application of on-line SFE-GC for the qualitative characterization of orange peel is shown in Figure 8. Approximately 100 mg of orange peel was extracted for 10 min intervals at two progressively higher pressures using supercritical carbon dioxide at 40 °C. After each extraction a temperature programmed GC analysis was conducted. In this particular case, the extractions were discontinued during the GC analyses, but if more selective fractionations had been desired they could have been continued during the analyses and the effluent directed to the collection vessel. Since orange peel was expected to contain volatile flavor compounds, a chromatographic column with a thick stationary phase film and subambient collection and focusing of the extraction effluent were used in the GC analysis. The chromatogram obtained at the lower extraction pressure has relatively higher concentrations of the earlier eluting compounds and the higher pressure extraction has relatively higher concentrations of the later eluting compounds. Since the entire range of compounds from the orange peel exhibited substantial solubility in the carbon dioxide at the low pressure, it was not possible to obtain a highly selective extraction. Utilization of a fluid having lower solvating power, such as SF_6, Xe, or Kr (17), would provide greater potential for selective fractionation of volatile and semivolatile low molecular weight compounds. This example illustrates the potential of SFE-GC of providing nearly automated sample preparation and analysis for the qualitative characterization of sample matrices. More selective SFE-GC analyses of higher molecular weight compounds with lower carbon dioxide solubilities have been reported elsewhere (13).

Quantitative Extraction Analyses. Exhaustive extraction of a sample matrix at a pressure where all of the components of interest are soluble provides the capability for a quantitative analysis. This method is illustrated with the example shown in Figure 9. XAD-2

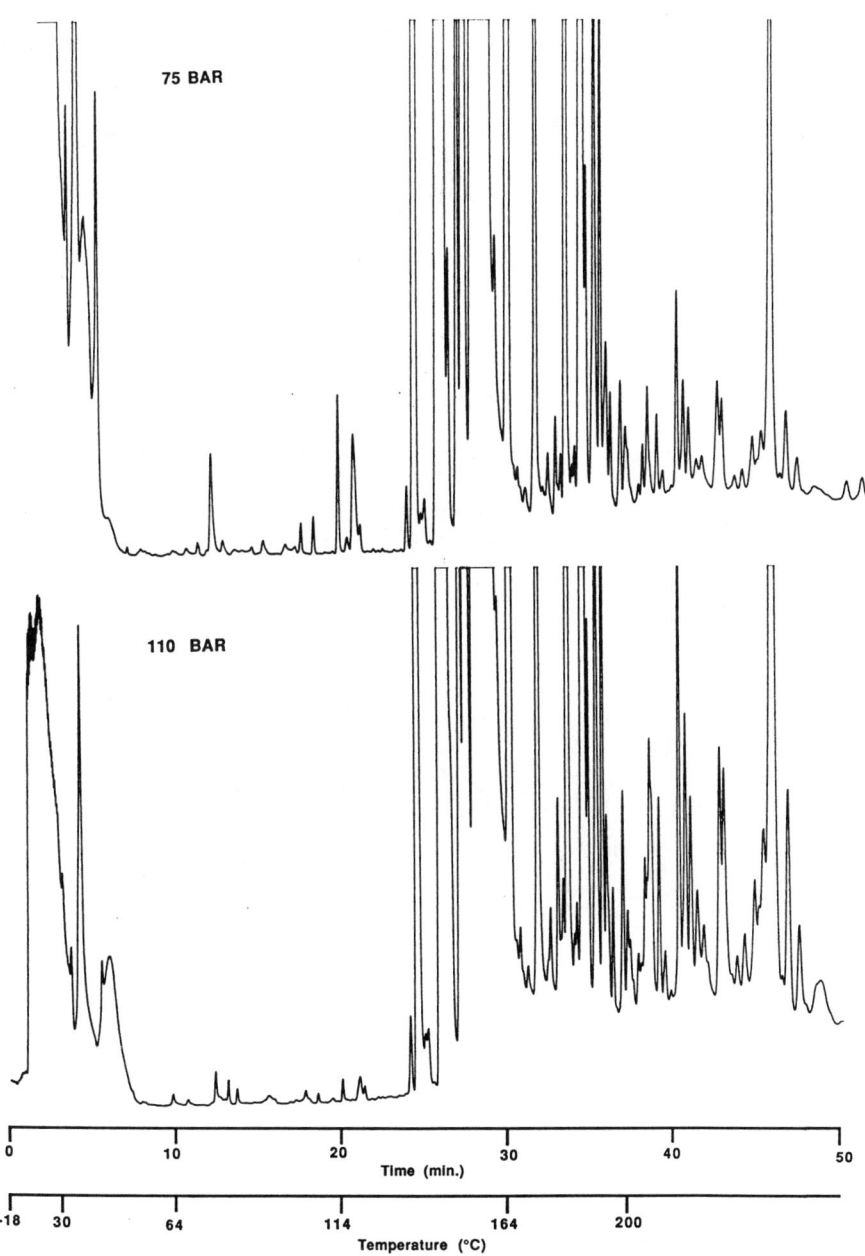

Figure 8. Capillary GC chromatograms obtained from supercritical carbon dioxide extraction at two different pressures of orange peel.

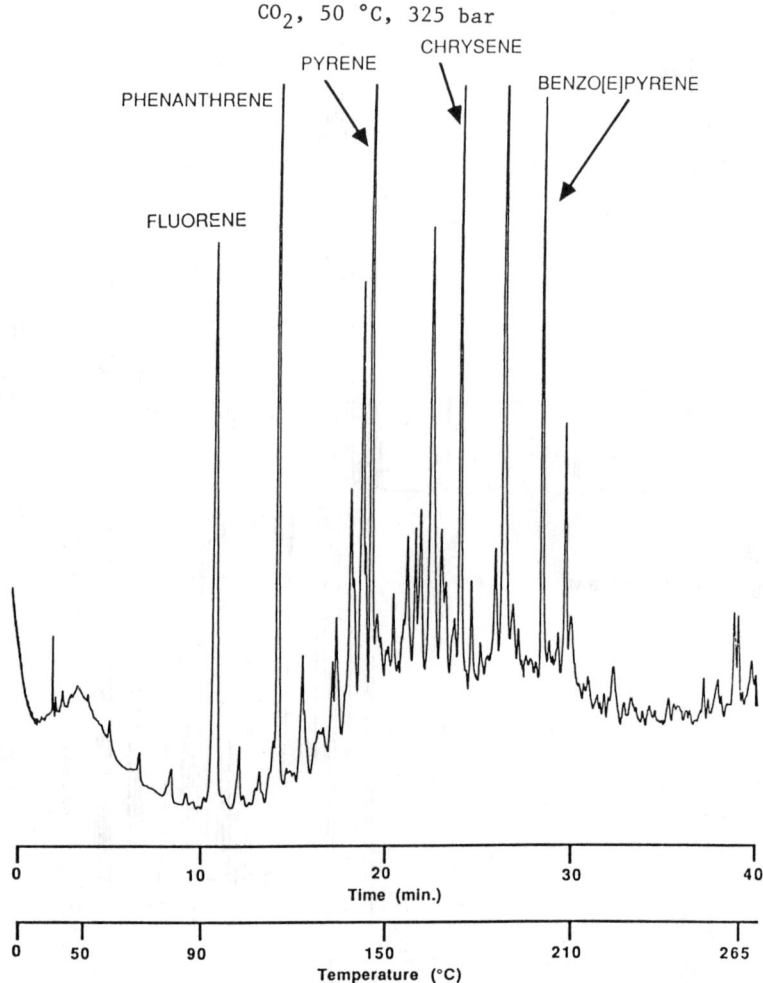

Figure 9. Capillary GC chromatogram obtained from the supercritical carbon dioxide extraction of XAD-2 resin.

resin that had been spiked with 7.5 ng each of several polycyclic aromatic hydrocarbons (PAH) was extracted with carbon dioxide at 325 bar and 50 °C for 7 min. The flow rate of decompressed carbon dioxide was over 250 mL/min, which allowed a sufficiently large fluid volume to be used to obtain an exhaustive extraction. The chromatographic oven was held at 0 °C during the extraction to aid focusing of the analytes and was then temperature programmed after the extraction was completed to effect the chromatographic separation. The detector response was calibrated for each compound immediately prior to the extraction-analysis. The specific compounds used, the spike levels, and the recovered quantities of each compound are listed in Table IV. Within experimental error (~15%), quantitative extraction and recovery of all the compounds, except benzo[e]pyrene, was obtained. Benzo[e]pyrene has lower solubility in carbon dioxide and was not completely extracted with the fluid

Table IV. Recovery of PAH from XAD-2 Resin

Compound	Spike Level (ng)	Recovered Amount (ng)	% Recovery
Fluorene	7.6	7.6	100
Phenanthrene	7.4	7.1	96
Pyrene	7.5	6.6	88
Chrysene	7.3	6.4	88
Benzo[e]pyrene	7.4	4.9	66

volume utilized. However, the remaining benzo[e]pyrene was recovered with a second extraction. The extraneous peaks in the chromatogram can be attributed to impurities concentrated from the carbon dioxide or from contaminants extracted from the XAD-2. Since several milliliters of liquid carbon dioxide are decompressed through the chromatographic column with quantitative extractions, it is essential to have very pure carbon dioxide to prevent serious contamination.

On-line SFE-GC has also been used as an alternative to thermal desorption gas chromatography of Tenax sampling devices for quantitative analysis of volatile organic compounds (VOC). The SFE-GC approach allows the analytes to be recovered from the adsorbent sampling devices at mild temperatures, which prevents thermal decomposition and other problems associated with high temperature desorption. This work is described in detail elsewhere (Wright,B.W.; Kopriva, A.J.; Smith, R.D. submitted for publication).

Conclusions

Off-line supercritical fluid extraction, simultaneous ultrasonic supercritical fluid extraction, and on-line supercritical fluid extraction-gas chromatography have been described. These analytical supercritical fluid extraction methods provide the potential for very rapid extraction rates and compatibility with on-line analytical methods. Extraction rate increases of over an order of magnitude compared to Soxhlet methods have been demonstrated and even greater increases seem feasible. Optimization of fluid solvating conditions also provides the potential for selective fractionation of specific analytes. The application of ultrasound during supercritical fluid extraction provides an efficient mechanism of stirring and may enhance extraction rates in macroporous materials by inducing convection through the pores. On-line extraction-analysis methods combine sample preparation and analysis and provide the potential for rapid and highly sensitive analyses. Additional studies and further development of analytical SFE methods are needed for a more complete evaluation and to allow achievement of their full potential.

Acknowledgments

Although the research described in this article has been funded wholly or in part by the United States Environmental Protection Agency through Interagency Agreement DW 899930650-01 through a Related Services Contract with the U.S. Department of Energy under Contract DE-AC06-76RLO 1830, it has not been subjected to Agency review and therefore does not necessarily reflect the views of the Agency and no official endorsement should be inferred. Mention of trade names or commercial products does not constitute endorsement or recommendation for use.

Literature Cited

1. Stahl, G.M.; Wilke, G. Extraction with Supercritical Gases; Verlag Chemie: Deerfield Beach, FL, 1980.
2. Paulaitis, M.E.; Krukonis, V.J.; Kurnick, R.T.; Reid, R.C. Rev. Chem. Eng. 1983, 1, 179-250.
3. McHugh, M.A.; Krukonis, V.J. Supercritical Fluid Extraction, Principles and Practice; Butterworths: Boston, MA, 1986.
4. Unger, K.K.; Roumeliotis, P.J. J. Chromatogr. 1983, 282, 519-526.
5. Smith, R.D.; Udseth, H.R.; Wright, B.W. In Supercritical Fluid Technology; Penninger, J.M.L.; Radosz, M.; McHugh, M.A.; Krukonis, V.J., Eds.; Elsevier: Amsterdam, The Netherlands, 1985; pp 191-223.
6. Sugiyama, K.; Saito, M.; Hondo, T.; Senda, A.J. J. Chromatogr. 1985, 332, 107-116.
7. Hawthorne, S.B.; Miller, D.J. J. Chromatogr. Sci. 1986, 24, 258-264.
8. Schantz, M.M.; Chestler, S.N. J.Chromatogr. 1986, 363, 397-401.
9. Kalinoski, H.T.; Udseth, H.R.; Wright, B.W.; Smith, R.D. Anal. Chem. 1986, 58, 2124-2129.
10. Campbell, R.M.; Lee, M.L. Anal. Chem. 1986, 58, 2247-2251.
11. Capriel, P.; Haisch, A.; Khan, S.U. J. Agric. Food Chem. 1986, 34, 70-73.
12. Wright, B.W.; Wright, C.W.; Gale, R.W.; Smith, R.D. Anal. Chem. 1987, 59, 38-44.
13. Wright, B.W.; Frye, S.R.; McMinn, D.G.; Smith, R.D. Anal. Chem. 1987, 59, 640-644.
14. Hawthorne, S.B.; Miller, D.J. Anal. Chem. 1987, 59, 1705-1708.
15. Fogler, H.S. In Sonochemical Engineering; Fogler, H.S., Ed.; AIChE: New York, 1971; Vol.67, p 1.
16. Chynoweth, A.G.; Schneider, W.G. J. Chem. Phys. 1951, 19, 1566-1569.
17. Smith, R.D.; Frye, S.L.; Yonker, C.R.; Gale, R.W. J.Phys. Chem. 1987, 91, 3059-3062.

RECEIVED November 13, 1987

Chapter 4

Supercritical Fluid–Adsorbate–Adsorbent Systems

Characterization and Utilization in Vegetable Oil Extraction Studies

Jerry W. King, Robert L. Eissler, and John P. Friedrich

Northern Regional Research Center, Agricultural Research Service, U.S. Department of Agriculture, Peoria, IL 61604

> Supercritical fluid (SCF) extraction of vegetable oils from a seed matrix may require the incorporation of an adsorbent bed to remove odoriferous compounds from the recycle gas stream. In this study, an elution pulse chromatographic technique has been developed which allows the determination of breakthrough volumes (BTV) for an array of odor-causing sorbates on Tenax-TA, XAD resins, and activated carbon adsorbents. Data taken over a 100-400 atmosphere range at three different temperatures indicate that the adsorbent retention capacity is rapidly lost at pressures beyond 200 atmospheres. In specific cases with the synthetic polymer sorbents, the supercritical carbon dioxide changes the sorption capacity of the resin phase by altering the physical morphology of the sorbent. Differential heat of adsorption measurements suggest that modification of the gas-solid interface by the supercritical fluid is enhancing migration of the sorbates through the column bed. Calculations incorporating the derived BTV data base illustrate how conditions can be altered to improve trapping efficiency by varying the fluid pressure, temperature, flow rate, and sorbent charge.

Adsorption as a complementary process to supercritical fluid extraction confers an extra degree of flexibility in segregating and fractionating solutes dissolved in the fluid phase. Literature citations of adsorption coupled with supercritical fluid extraction range from the "classical" caffeine extraction with CO_2 (1) to recent attempts to separate cholesterol from butter using adsorbent beds of charcoal and silica gel (2). Prior studies utilizing adsorbents in the presence of supercritical fluid media have been reviewed by King (3), who has commented on the lack of fundamental knowledge on adsorbate(sorbate)/adsorbent(sorbent)/supercritical fluid systems. Indeed, with the exception of the sorbent regeneration studies performed at Critical Fluid Systems in the last decade (4,5), there is a paucity of data on how industrially useful compounds behave in a supercritical fluid process stream passing through a sorbent column.

The determination of adsorption isotherms, adsorption coefficients in the Henry's Law region, heats of adsorption, along with the breakthrough characteristics of sorbates provide a basis for enlarging our theoretical understanding of the adsorption process and contribute to the design of adsorbent units operating under high pressure. Data such as the breakthrough volume characteristics of a sorbate partitioned between an active sorbent and a compressed fluid phase allow the prediction of the service lifetime of the sorbent bed and define conditions which are appropriate for selectively retaining solutes from the gas phase after the primary extraction step. More recently, the utilization of supercritical gases to selectively desorb analytes from sorbent resins, such as Tenax, XAD-2, and polyurethane foams (6-8) suggest that additional research is needed to specify conditions which are optimal for the rapid removal of analytes from these purified sorbents. Knowledge of the breakthrough characteristics for given sorbate/sorbent pairs permits the specification of sampling protocols for trapping desired components, and conversely, allows estimates to be made of the conditions required for effective desorption of reversibly-adsorbed components. As shown in an earlier study (3), the breakthrough volume asymptotically approaches a limiting volume defined by the void volume of the adsorbent bed as the system pressure is increased, hence there is a discrete pressure range over which the adsorbent bed can be used for isolating or fractionating components which are solubilized in the supercritical fluid phase.

The present work was initiated to measure the breakthrough volumes for selected sorbates which can potentially contaminate the recycle gas stream in a continuous supercritical fluid extraction system designed for processing vegetable oils. Such components, if not selectively adsorbed, can recontaminate the residual protein meal as well as extracted oil, and contribute to inferior flavor test panel scores on both products. The data base accumulated in this study also offers guidelines for designing an appropriate off-stream analytical sampling device, based on sorbent resin methodology, for collecting and characterizing the odoriferous components in the recycled supercritical fluid stream.

A schematic of the supercritical fluid extraction system is shown in Figure 1. The system is divided into two separate pressure regions; one section usually operating at pressures between 10-12,000 psig and 80°C for extracting the oil from the seed bed and a second section between the back pressure regulator and compressor held at 2500 psig and 80°. Within this lower pressure section, the oil is precipitated from the supercritical fluid phase and the solvent gas recycled back to the compressor over a bed of activated carbon. Details of the extraction conditions and equipment can be found in earlier publications (9,10). Although the system in Figure 1 was expressly designed for the extraction of vegetable oils, it is fairly typical of many supercritical fluid extraction systems, thus the data and

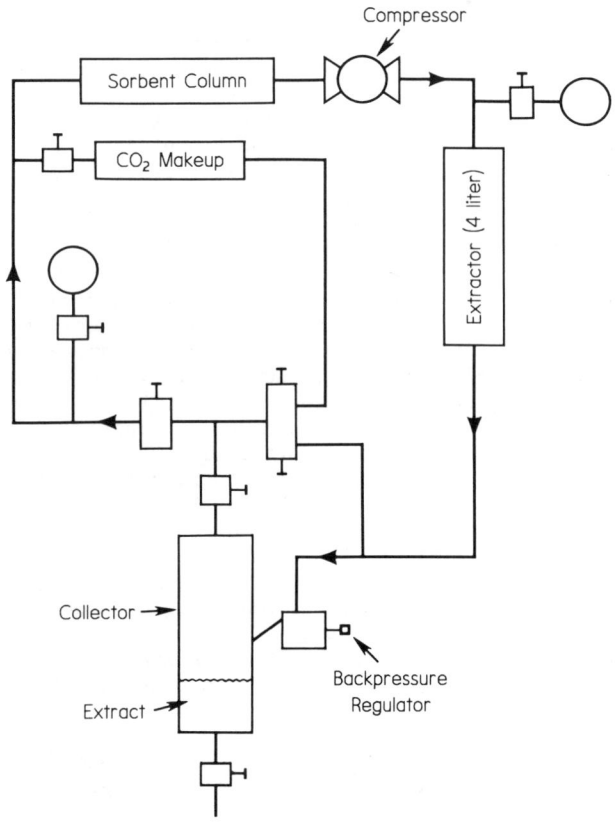

Figure 1. Continuous recycle SCF-CO_2 extractor.

techniques described in this paper can be extrapolated to other processing situations as required.

The technique used to measure the breakthrough volumes is similar to that reported by King (3) in which a modified chromatographic apparatus is used in the elution pulse mode to measure peak maximum retention volumes as a function of gas pressure. As described previously, trace contaminents in the SCF mobile phase elute as symmetrical (gaussian) peaks when adsorption occurs in the Henry's Law region of the sorption isotherm. The peak maximum corresponds to the 50% breakthrough volume of a sigmoidal sorbate wavefront as the adsorbent bed begins to saturate with respect to the challenge concentration of the respective sorbate. This particular quasi-equilibrium retention volume is invariant with respect to the sorbate fluid concentration and column generated non-equilibrium contributions to peak spreading, and hence is a preferred and accurate measure of the breakthrough characteristics of a given sorbate/sorbent combination (11,12). Such an approach has been utilized by many investigators to determine breakthrough data required in environmental sampling (13-15), industrial hygiene monitoring (16,17), and flavor/odor characterization (18). Good agreement between the measured retention volumes and experimental breakthrough curves has been reported (11).

The advantage of the pulse chromatographic method lies in its experimental simplicity and the ability to rapidly accumulate data which characterizes the retentive behavior of adsorbates in the presence of a supercritical fluid phase. By incorporating an apparatus which is capable of duplicating the process conditions under which the adsorbent unit is operating, one can estimate the service lifetime of the adsorption column as a function of pressure and temperature. Although retention volume data alone cannot be used to elucidate the relative contribution of competing mechanisms to the migration enhancement of sorbates in an adsorption column, such information serves a practical and applied purpose as illustrated in this study. Given the current state of our understanding of supercritical fluid/adsorbate/adsorbent systems, experimental measurement of the governing physicochemical parameters may be the most efficient and reliable way of operating and designing systems for commercial use.

Experimental

A Hewlett Packard Model 1082B supercritical fluid chromatograph was utilized for the measurement of sorbate retention volumes in this study. The instrument is a modified liquid chromatograph consisting of cooled pump heads to facilitate accurate metering of liquefied carbon dioxide from a reciprocating diaphragm pump, an ultraviolet high pressure detector flow cell capable of withstanding pressures to 400 atmospheres, and a precise back pressure regulator to maintain pressure during the chromatographic analysis. This instrumental design permitted the precise determination of BTV without substantial modification to the commercial instrument. Experimental data using supercritical

CO_2 as a carrier fluid were collected over a pressure interval of 100-400 atmospheres in a temperature range of 35-90°C. Tests performed at flow rates ranging from 0.5 to 5.0 ml/min showed that the peak maximum retention volume was constant with respect to flow rate, thereby permitting extrapolation of the measured retention volumes to higher carrier velocities.

Detection of the sorbate peak after elution from the adsorption column was accomplished using a variable wavelength detector. With the exception of 2,4-decadienal, many of the solutes utilized in this research had small molar absorptivities and could only be detected in the far ultraviolet region of the spectrum. Convenient wavelengths for monitoring purposes were 215-230 nm for the carbonyl-bearing compounds and 200 nm for the n-alcohols. Injection volumes of 1.0 µL or less of the neat solute were sufficient for detection in the reported studies.

Sorbent columns fabricated within our laboratory consisted of 316 stainless steel tubes, 3 to 12 inches in length, having an internal diameter ranging from 5/16" (0.794 cm) to 7/32" (0.556 cm). The dimensions of these columns permitted adjustment of the sorbent weight to between 0.5 - 3.0 grams, and hence the retention time range that was recorded experimentally. Adsorbent weight was determined on an analytical balance both before and after the chromatographic experiments. Reproducible sorbent weights after experimental use could only be recorded after allowing overnight outgassing of the sorbent column prior to weighing.

Table I summarizes the adsorption selection incorporated into the experimental studies. Four pre-purified synthetic resin sorbents obtained from Alltech Associates (Deerfield, IL) were used along with an activated carbon sample (Union Carbide Grade 6GC) procured from our pilot plant. The coarse particle size of the selected adsorbents in conjunction with the column dimensions and experimental flow rates yielded very small pressure drops (< 2 atm) across the sorbent bed, as ascertained by inlet and outlet pressure gauges before and after the column, respectively. Because of this small pressure gradient within the sorbent column, the recorded experimental breakthrough data were taken essentially at isobaric conditions.

Table I. Adsorbent Properties

Adsorbent	Mesh Size	Surface Area	Structure
Tenax-TA	20-35	35 m^2/g	diphenyl phenylene oxide
XAD-2	20-50	300 m^2/g	styrene/divinylbenzene
XAD-7	20-50	450 m^2/g	acrylic ester
XAD-8	20-50	150 m^2/g	acrylic
Activated carbon	10-24	>1000 m^2/g	carbon

The sorbates utilized in this study were chosen by consulting the literature on the odor/flavor characteristics of soybean oil and meal (19,20). Specific components, such as 2,4-decadienal, 2-pentylfuran, and ethyl esters have been identified as major contributors to the flavor chemistry of soybean oil by Frankel (22) and other investigators (23). These compounds and selected solutes comprising homologous series of 2-methyl ketones, aliphatic aldehydes, and n-alcohols were chosen as sorbates based upon their probable occurrence in trace quantities during the supercritical fluid process. It should be noted that many of the sorbates used in this study exhibit appreciable volatility and would be expected to have small breakthrough volumes on the sorbents cited previously.

Several tests were performed to assure that the measured retention volumes reflected sorbate retention in the miniature adsorption column. Thermocouple-based measurements of the injection valve compartment (which is mounted outside the column oven) revealed a 2°C lag in temperature from that recorded in the column oven. This small temperature difference had a negligible effect on the solute retention volume at flow rates of 0.5 ml/min and is further minimized at higher carrier gas flow rates due to prior thermal equilibration of the gas in the column oven. The contribution of instrument dead volume was assessed by measuring the retention volume of a test solute (methanol) in the chromatographic system in the absence of the adsorption column. The system dead volume was found to be less than 0.1 ml, a negligible contribution to the measured sorbate retention volumes.

Collection of the retention data was initially taken on Tenax and XAD-2 adsorbents at 150, 250, and 350 atmospheres. As experimental work progressed, additional data was taken at closer pressure intervals to better define the trend in breakthrough volume with fluid pressure. Measurement of retention volumes below 100 atmospheres was difficult, due to the "threshold pressure" solubility limit (as defined by the sensitivity of the UV detector) of the test solute probes (23). Most of the generated retention data were measured at three temperatures: 40, 60 and 80°C.

Retention volume data accumulated over a period of two weeks on a particular sorbent showed excellent reproducibility. Retention volumes could also be reproduced at a given pressure regardless of whether the data was taken by increasing or decreasing the pressure between intervals. No bias in the collected data as a function of temperature could be detected regardless of whether the measurements were made by increasing or decreasing the temperature. Multiple determinations of the specific BTV for a given sorbate on the same column yielded a standard deviation of $\pm$ 0.3 mL/gram. Prolonged use of the Tenax columns resulted in a gradual reduction in BTV for a given sorbate at a specific pressure. The reasons for this reduction in BTV will be discussed in the next section.

Calculation of the breakthrough volume is made by using the simple equation:

$$BTV = (F)(t_r) \qquad (1)$$

where F = flow rate of the supercritical fluid at column temperature and pressure

t_r = time from injection to the elution peak maximum

For purposes of comparing the relative retentive properties of adsorbents, a specific breakthrough volume may be calculated by dividing the weight of the adsorbent into the above-defined volume. Equation 1 is relatively simple compared to the more elaborate calculations used to determine the net and specific retention volumes in thermodynamic gas chromatographic studies (24). This in part is due to the experimental design used and the desire to have data which reflect breakthrough characteristics under supercritical fluid conditions. Since the combination of liquid metering pumps and back pressure regulation permit independent control of fluid flow rate and pressure in the supercritical fluid state, there is no need to convert the measured retention volumes to actual column conditions or a standard temperature and pressure. The breakthrough volume measurements in this study have not been corrected for the column void volume, since over the higher pressure range, the net retention volume would be effectively zero in many cases. For this reason and the accompanying observation that certain synthetic adsorbents were physically changed when exposed to supercritical fluid CO_2, we have elected to present retention data uncorrected for column void volume to aid in the interpretation of the experimental data. In addition, from the perspective of sorbate holdup in the sorption column, a solute may be retained on the adsorbent bed if the sampling period is brief, even though it spends all of its residence time in the supercritical fluid phase.

Results

Our initial studies focused on the measurement of breakthrough volumes for a variety of sorbates on Tenax-TA resin, a sorbent which can be employed in thermal desorption methods for characterizing the trace components in the gas stream. It was anticipated that the trend in sorbate breakthrough volume with pressure would parallel earlier findings (3) in which the retention volume decreased substantially at lower pressures while attaining a constant value in the higher pressure regime. However, as shown in Figure 2, the BTV determination at 250 atmospheres and 40°C was approximately twice as large as those recorded at 150 and 350 atmospheres for all of the sorbates examined. Additional data collected over the 100-400 atmosphere range substantiated the initial findings, indicating that there was a distinct BTV maximum occurring in the 200-250 atmosphere region (as typified by 2,4-decadienal).

Close inspection of Figure 2 indicates a slight fractionation effect between the most non-volatile component (2,4-decadienal) and the other sorbates, particularly in the lower pressure region (100-150 atm). This "light/heavy" selectivity trend has been observed in other gas-solid chromatography studies, particularly for solutes comprising a homologous series (25). For the homologues examined in this research, little individual separation was observed beyond 150 atmospheres. In the process situation cited previously or for analytical collection of sorbates, separation of the individual sorbates is not required.

To further confirm and elucidate the cause of the erratic pattern in retention behavior shown in Figure 2, breakthrough volumes were determined for thirteen compounds at three different pressures on the Tenax resin. As shown in Figure 3, the individual sorbates tended to cluster around a discrete BTV at each individual pressure, suggesting that the retention volume trend was independent of the sorbate type. This observation coupled with the substantial volatility enhancement factors expected and recorded for similar solutes (26) in $SCF-CO_2$ support the contention that the sorbates are passing through the adsorption column with retention volumes equivalent to the void volume of the column bed. Such results suggest that this reproducible retention trend for so many solutes may be related to changes in the sorbent structure.

To test whether the retention trends on Tenax were reproducible and invariant with respect to experimental protocol, several select sorbates were chromatographed over several days at three different temperatures. As shown for ethyl caproate in Figure 4, the pattern is reproducible for data taken 2-3 days apart at 40°C. Cycling the sorbent bed while taking individual retention volume measurements from low to high pressures and back (as indicated by the different symbols in Figure 4) produced no measurable change in retention volume pattern. This confirms that the three-fold difference in breakthrough volume in going from 150 to 250 atmospheres is outside the limit of experimental error. The trend in BTV with pressure for sorbates on Tenax as typified by ethyl caproate was also observed at 60 and 80°C. Accumulation of additional data points in the pressure intervals between 100-150 and 175-300 atmospheres showed that the ethyl caproate retention minimum occurred between 100-120 atmospheres as indicated in Figure 5. The BTV maximum was once again found to occur between 200-250 atmospheres for ethyl caproate and other solutes, with a second BTV minimum occurring at about 275 atmospheres.

The results obtained in Figures 2, 4, and 5 can be explained by considering the individual factors which influence the breakthrough volume trends over discrete pressure ranges. Initial decreases in the retention volume are due to the enhanced solvation of the sorbate (solute) in the supercritical fluid phase abetted by displacement of the sorbate from the sorbent surface by CO_2 as the pressure is increased. This surface effect has been documented previously (3) and recent breakthrough curves studies by Groninger (27) for light gas components in compressed methane on molecular sieve confirm the competitive displacement effect. Assessment of the relative contributions of two reinforcing mechanisms responsible for the reduction in BTV would require data from independent solubility and adsorption experiments.

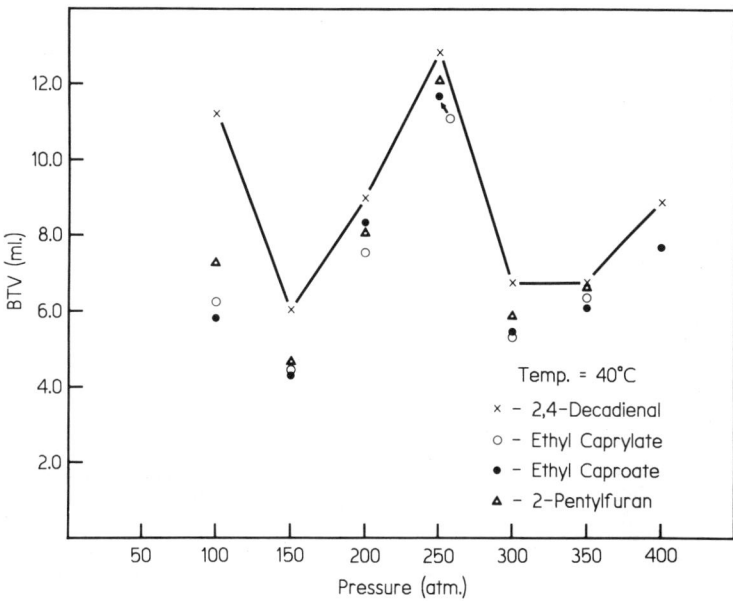

Figure 2. BTV pattern as a function of pressure on Tenax.

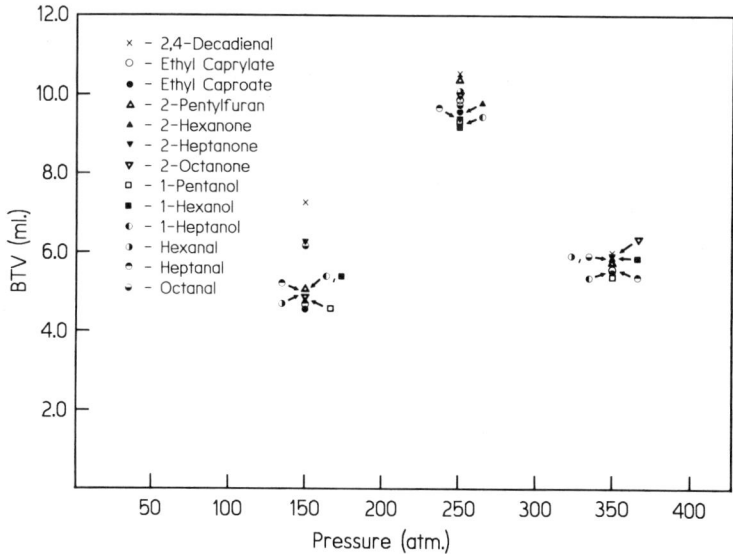

Figure 3. BTV for a large number of adsorbates versus pressure on Tenax at 60°C.

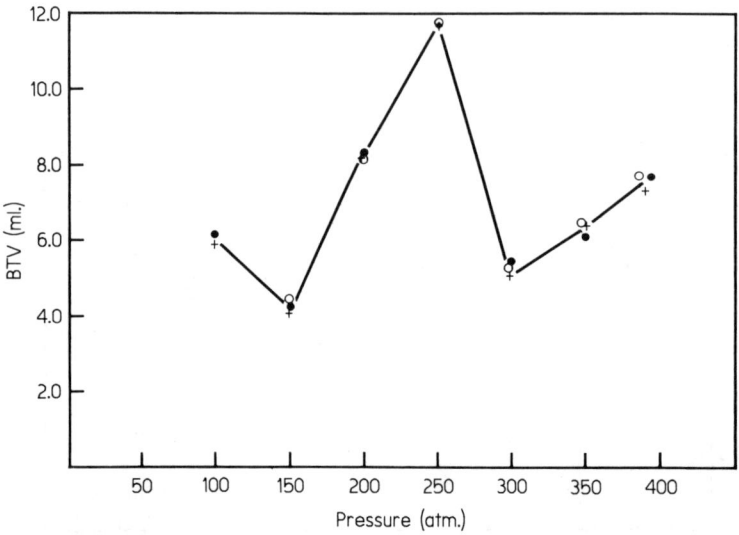

Figure 4. BTV for ethyl caproate as a function of pressure on Tenax at 40°C.

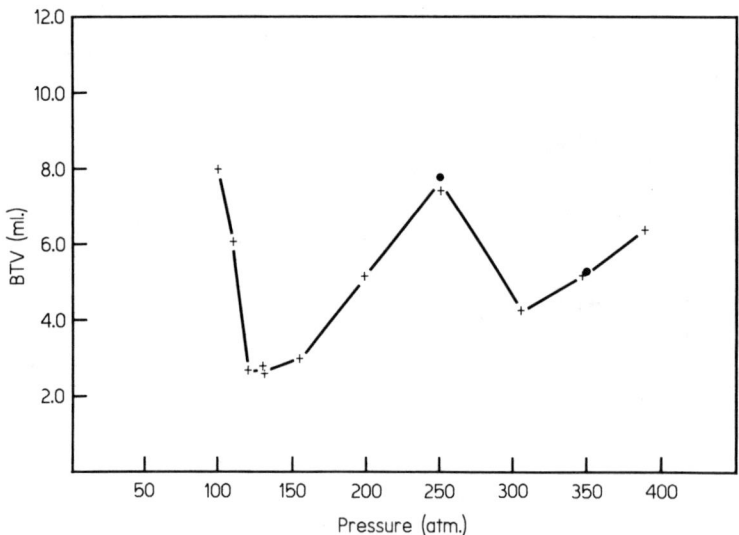

Figure 5. BTV for ethyl caproate as a function of pressure on Tenax at 80 °C.

The increase in BTV at intermediate pressures for the sorbate/Tenax systems is primarily due to morphological changes in the synthetic resin sorbent. Increases in column void volume reflect a change in the internal porosity of the resin which will increase the available surface area for sorbate/sorbent interaction. Carbon dioxide has been shown to induce swelling in coal leading to higher recorded surface areas for the coal matrix (28). It is also well known that many elastomers in compressed gases (29) as well as cross-linked resins in liquid solvents (30) will swell appreciably depending on the magnitude of the solvent (SCF)-polymer interaction. Recent studies on polymethylmethacrylate polymer (31) show that dimensional changes equivalent to 16% of the polymer's original length can be affected by imbibing the polymer lattice with compressed CO_2. Plastization of similar polymers by compressed gases can also cause the glass transition temperature, T_g, of the bulk polymer to be lowered (32), resulting in physical deformation of the resin. It should be noted that the solubility parameter of CO_2 at 250 atmospheres and 40°C is 7.7 $cal^{1/2}/cm^{3/2}$, a value equivalent to many liquid hydrocarbon solvents commonly used in polymer swelling studies.

At pressures approaching 300 atmospheres, the BTV for sorbates on Tenax again decrease inferring that the dense gaseous environment surrounding the resin is less capable of inducing swelling in the polymer matrix. A similar trend in chromatographic measured void volumes has been noted by Novotny (31,34) for helium injections into a SCF-CO_2/surface-bonded silicone polymer capillary column system. Chromatographic void volume measurements may offer an alternative method to assess solvent-induced polymer swelling, provided the polymer is not miscible with the supercritical fluid.

Similar measurements to those discussed above were performed on a cross-linked styrene/divinylbenzene polymer, XAD-2, using selected sorbates. The breakthrough volume trend at 40°C for four of the sorbates is shown in Figure 6. As in previous figures, the trend in ethyl caproate BTV is emphasized by connecting the data points at each consecutive pressure by a straight line. On the XAD-2 resin, ethyl caproate does not show the initial decrease in BTV between 100-150 atmospheres as recorded for sorption on Tenax. However, in the intermediate pressure range (150-300 atm), the BTV for ethyl caproate increases to twice that value found in the lower pressure regime. Experimental data taken at 150, 250, and 350 atmospheres for ethyl caprylate and 2-pentylfuran indicate that these sorbates follow a similar pattern to ethyl caproate. The aldehyde sorbate, 2,4-decadienal shows a gradual reduction in BTV with fluid pressure. Such complex retention patterns would be difficult to predict and show the value of the experimental method in determining BTV. These results also indicate that in some cases, resinous adsorbents may yield enhanced retention capacity at higher pressures.

As with the Tenax system, there is evidence here that the sorbent may be perturbed by the supercritical fluid. Wang and coworkers (35) have shown that pressurized CO_2 can severely plasticize polymers. Creep tests (36) indicate that appreciable changes in bulk polymer moduli as well as differential changes in the glass transition temperature (up to 50 to 70°C) occur at pressures under 100 atmospheres in polystyrene. Additional dilation data for carbon dioxide absorbing into polystyrene (37) support the above study and indicate that the superimposition of a polymer phase change may also influence the retentive capacity of the polymeric sorbent.

Results obtained for the same sorbates on XAD-2 at 80°C are depicted in Figure 7. In this case the sorbate retention behavior coincides with retention volume trends reported by numerous investigators in gas-solid chromatography (3). Unlike the sorbate trends recorded at 40°C over the same fluid compression range, the BTV of the compounds decrease appreciably at pressures below 200 atmospheres and approach a constant value in the higher pressure region. As illustrated, ethyl caproate shows a 4-5 fold reduction in BTV over a 100 atmosphere interval and the selectivity and retentive capacity of the sorbent for the other sorbate moieties is eliminated at pressures exceeding 200 atmospheres. Note that there is still some retention toward the sorbates at 170 atmospheres, a condition corresponding to the low pressure side of the oil extraction system. The breakthrough volume pattern for sorbates on XAD-7 and 8 at 80°C is very similar to that observed for the XAD-2 resin and a finite adsorption capacity exists at pressures below 200 atmospheres for the sorbates studied.

When adsorption studies were conducted over a prolonged time period (one month or more) on the same resin column, it was observed that a gradual reduction in the BTV occurred corresponding to 10-20% of the initial BTV value. This phenomenon was independent of the nature of the adsorbate and was particularly obvious in the intermediate pressure ranges when using Tenax as an adsorbent. The potential for resin comminution, sintering, or other physical changes in sorbent structure due to repeated cycling from low to high pressures over the duration of the experimental work could account for the above observations. Consequently, a scanning electron microscopy (SEM) study was undertaken to compare the virgin adsorbent structure with the resins exposed to supercritical fluid carbon dioxide. Electron photomicrographs of unexposed Tenax and SCF-exposed resin are shown in Figures 8a and 8b, respectively. At a magnification level of 17X, there appears to be no difference between the irregular particles of Tenax resin. Scanning over a large number of particles at an increased magnification level, 550X, revealed differences in the porosity of the virgin and exposed adsorbent. High magnification scans at 4000X (Figure 9a and 9b) showed that the unexposed Tenax had a finer porous structure while the used sorbent had less internal porosity and was morphologically different from the virgin specimen. The condensed structure of the exposed Tenax resin might have been

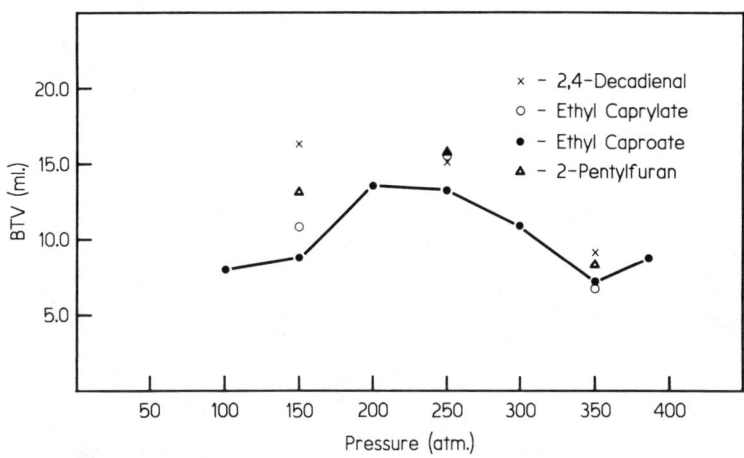

Figure 6. BTV for various sorbates on XAD-2 resin at 40°C.

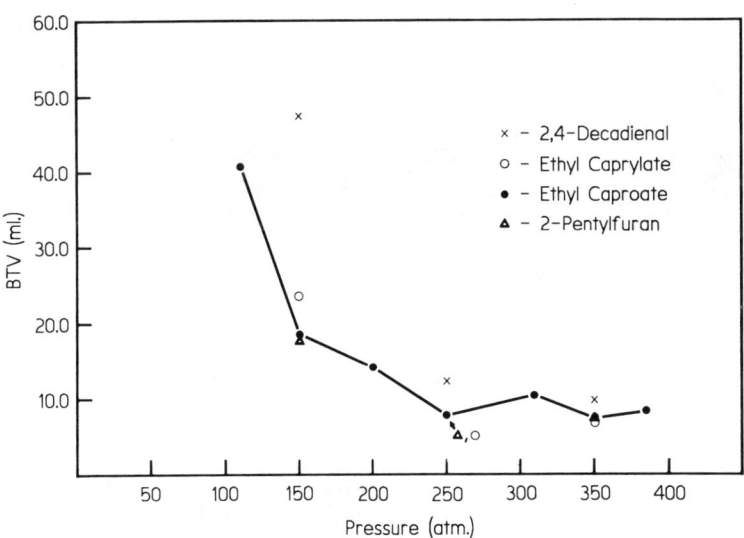

Figure 7. BTV for various sorbates on XAD-2 resin at 80°C.

Figure 8. SEM photographs of Tenax resin (17X), (a) virgin, (b) SCF-exposed.

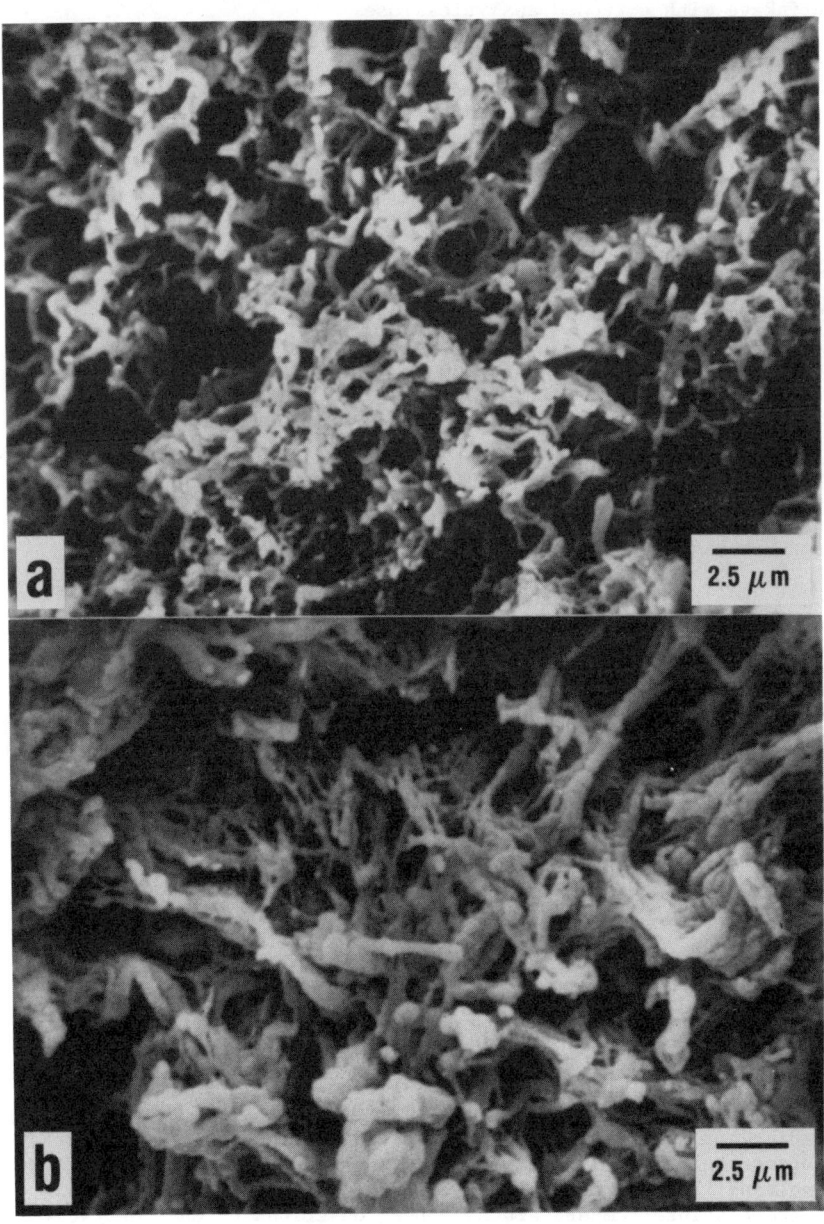

Figure 9. SEM photographs of Tenax resin (4000X), (a)virgin, (b)SCF-exposed.

anticipated since chromatographic columns of the diphenyl phenylene oxide-based polymer have been shown to undergo bed shrinkage with extended analytical use.

A similar microscopic examination was also conducted on the XAD series of sorbents. Figures 10a and 10b show low magnification (27X) scans of the suspension polymerization-produced beads of the XAD-2 sorbent, which consist of aggregates of micro-particulates that comprise the spherical resin. The porosity of these adsorbents arises from the intraparticle voids between the microparticulates (38), which suggested that scans at increased magnification would be more informative. Photographs at 23,000X magnification are shown in Figures 11a and 11b for the virgin and SCF-treated sorbent, respectively. The clusters of microparticulates can be seen on the surface of the sorbent at this magnification, a pattern that was found to be consistent from bead to bead. The XAD-2 sample used for the extended BTV studies (Figure 11b) shows evidence of microparticulate sintering and an attendent loss of intraparticle porosity. The large fissures visible in Figure 11b are due to electron beam damage from the microscope's radiation source. Similar SEM studies on the XAD-7 and 8 sorbents revealed changes in their polymeric matrix morphology, however these pressure-induced alterations were observed to be less drastic due to the shorter experimental exposure times experienced by the acrylic resins.

The physical changes in adsorbent structure revealed by the scanning electron microscope studies have been observed with other polymeric materials exposed to supercritical fluids. McHugh and Krukonis (39) have demonstrated with the aid of SEM, that polypropylene preforms can be rendered porous by extraction with supercritical CO_2 and propylene. However, these investigators noted that SCF extraction of the preformed polymer sheets at higher pressures and temperatures deformed the polymer matrix leading to a fused appearance when examined by SEM and a concomitant decrease in porosity. Similar interaction between porous polyurethane foam and supercritical CO_2 reported by Smith and coworkers (7) lead to a physical alteration of the exposed sorbent. Such pressure-induced structural changes lead to a decrease in available surface area for sorbate adsorption.

Additional evidence of the role of supercritical fluids in mediating the chromatographic retention behavior of sorbates at the gas-solid interface can be gleaned from differential heat of adsorption measurements. Although only three temperatures were used in this investigation, Van't Hoff plots were constructed at pressures of 150, 250, and 350 atmospheres for the Tenax and XAD-2 sorbents. The relationships between the logarithm of the BTV and the reciprocal of absolute temperature proved to be complex and linearity was only observed for the Tenax/sorbate systems at 250 and 350 atmospheres. Non-linear behavior in plots involving the logarithm of the capacity ratio versus reciprocal of the absolute temperature have been described by Chester (40) and attributed to specific temperature regions in which the retention behavior is either gas or liquid chromatographic in

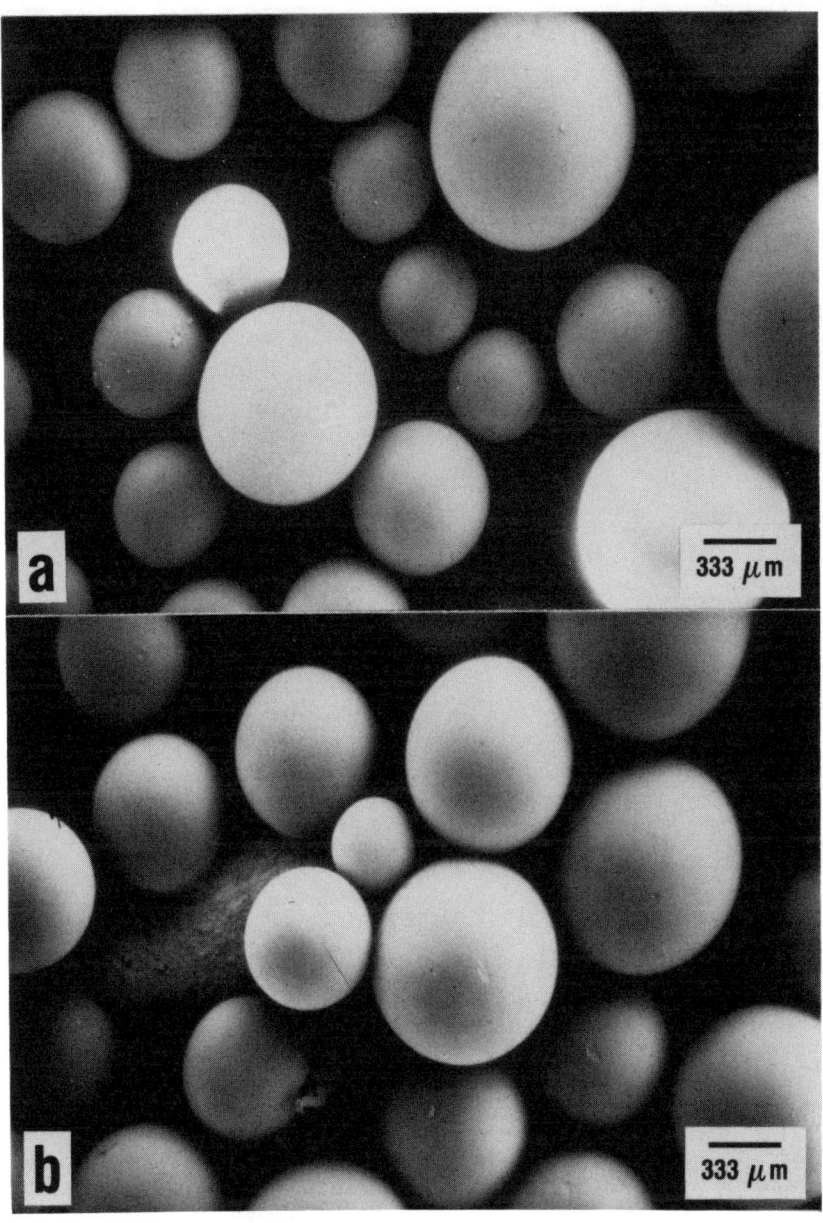

Figure 10. SEM photograph of XAD-2 resin (27X), (a)virgin, (b)SCF-exposed.

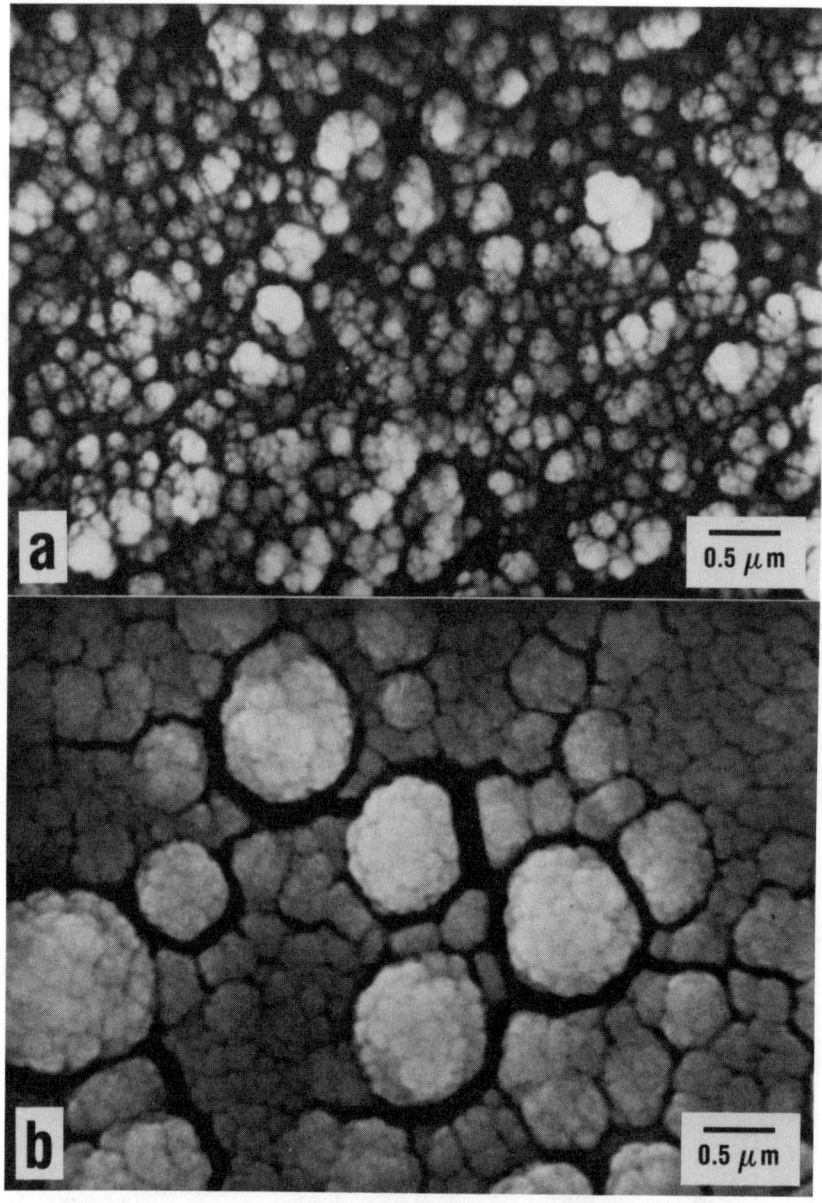

Figure 11. SEM photograph of XAD-2 resin (23,000X), (a)virgin, (b)SCF-exposed.

nature. Although the pressure range employed in this study is higher than that utilized in Chester's research, measurements at several other temperatures for specific solutes on Tenax at 150 atmospheres revealed a small slope change over the expanded temperature interval in the Van't Hoff plot. The Van't Hoff relationship for sorbates such as 2,4-decadienal on XAD-2 appeared to yield negative slopes at the 150 and 350 atmosphere compression levels, but changed sign from negative to positive as the pressure was decreased to 250 atmospheres. Such results may not be too surprising considering the complex array of factors which impact on the retention behavior of the sorbates. In the case of the temperature dependence of sorbates on XAD-2, the superimposition of a phase change may account for the recorded slope changes in the Van't Hoff plots. Inverse gas chromatographic studies conducted at low pressures on cross-linked styrene/divinylbenzene copolymer beads yield Van't Hoff plots with negative slopes over a rather large temperature range (41,42). As indicated previously, high pressure CO_2 can plasticize many polymers and reduce the glass transition temperature as much as 50 to 60°C. Such a pressure-induced phase change would place the T_g of the copolymer in the range of our experimental data. More experimental data needs to be taken as a function of temperature under isobaric conditions to confirm this inflection in the Van't Hoff plot.

The sorbate differential heats of adsorption per mole, $\Delta \overline{H}_A$, at higher pressures on Tenax were calculated using Equation 2:

$$\Delta \overline{H}_A = 2.303 \, R \, \frac{d \log (BTV)}{d (1/T)} \qquad (2)$$

where R = gas constant

T = absolute temperature (°K)

Excellent correlation coefficients (>0.98) were obtained for the 12 sorbates on Tenax and the average differential heat of adsorption for these compounds was -1.45 kcal/mole at 250 atmospheres and -0.87 kcal/mole at 350 atmospheres. The magnitude of the heats of adsorption indicate very weak enthalpic interactions between the adsorbate vapors and the adsorbent and the recorded values are significantly less than the corresponding heats of liquefaction of the adsorbates. The almost 0.6 kcal/mole difference in the average heat of adsorption recorded at 250 versus 350 atmospheres of pressure indicates weaker interaction between the sorbates and the adsorbent at the higher pressure. Such a result is consistent with surface modification of the interface by the supercritical fluid gas (the competitive adsorption effect). An interesting comparison is afforded between the measured differential heat of adsorption for 2-heptanone absorbing on Tenax from helium close to atmospheric pressure (11) and the corresponding values on Tenax from supercritical CO_2 at elevated pressures. The differential heat of adsorption for 2-heptanone at 250 atmospheres of CO_2 is -1.42 kcal/mole and -0.84 kcal/mole at 350 atmospheres. These are considerably below the value of -19.0 kcal/mole for adsorption from a low pressure helium stream and indicative of the strong modifying effect of supercritical CO_2 on the adsorption equilibrium. A similar conclusion can also be reached by comparing the heats of adsorption determined at the above pressures for n-alcohols on Tenax.

Experiments on activated carbon as an adsorbent proved difficult due to the strong adsorbate-adsorbent interactions, which in many cases led to diffuse, non-symmetrical elution profiles and lengthy breakthrough volumes. Examination of the elution profile at 80°C and a variety of pressures allowed a qualitative assessment to be made of adsorption characteristics of various compounds on the carbonaceous adsorbent. For example, injections of ethyl caproate and 2,4-decadienal on to the carbon column at 170 atmospheres (the process condition in our recycle extractor) showed no evidence of elution after 4-5 hours from the carbon bed. The sorbate, 2-pentylfuran broke through slowly yielding a diffuse symmetrical profile. Elevation of the pressure to 250 atmospheres resulted in a more rapid elution of the furan sorbate, slow desorption of the ethyl caproate, but no elution of the 2,4-decadienal. At 370 atmospheres, breakthrough for all of the above sorbates was instantaneous and the peak maximum retention volumes were very small. Even at 370 atmospheres, complete desorption of the sorbates required elution times of over one hour. Such a technique as the one described above may have merit in monitoring the desorption of contaminents from catalysts using supercritical fluids as has been reported in the literature (43).

The results obtained on activated carbon confirm its effectiveness for trapping volatile compounds at low pressures in the presence of supercritical fluid carbon dioxide. However, reversible recovery of the adsorbed species may prove difficult due to the extremely heterogeneous surface caused in part by the presence of oxygen-containing functional groups at the gas-solid interface (44). Similar conclusions have been reached by industrial researchers attempting to use supercritical fluids for the regeneration of adsorbents (45). Recovery problems also plague the use of activated carbon as an analytical sorbent material (18).

Discussion

The determination of breakthrough volumes provides a data base for estimating the service time of an adsorbent cartridge and guidelines for designing conditions to trap sorbates. To effectively use the experimental data; the breakthrough volumes must be expressed in volume per gram of adsorbent to correct for differences in the packing densities of the specific adsorbents. Although the conclusions from the graphical trends presented earlier remain the same, this method of presenting retention data allows for an intercomparison of the relative selectivity and trapping capabilities of candidate adsorbents. The breakthrough volume per gram also permits retention volume measurements to be extrapolated to larger or smaller sorbent weights to fit the particular process or experimental situation under consideration.

In Table II the breakthrough volumes determined by experimental measurement or extrapolation to 170 atmospheres and 80°C are expressed as milliliters per gram of adsorbent; the conditions corresponding to the low pressure side of the recycle extractor (Figure 1). An intercomparison of these values among the four adsorbents and six adsorbates is informative since the retention volumes represent the product of the available surface area of the adsorbent and the sorbate adsorption coefficients on the trapping medium (11). For most of the sorbent-sorbate combinations in Table II, sorbate retention on Tenax resin is less than on XAD series of resins. Such a difference would have been larger (three to fourfold in some cases) had the results not been expressed in per gram of adsorbent. When individual adsorbent packing densities are taken into account, the combined effect of thermodynamic selectivity and larger surface make the XAD sorbents the preferred medium for trapping the listed sorbates. It would also appear that the sorbent surface area is the prime factor influencing the retentive capacity of the various adsorbents despite their variation in chemical structure. For example, the relatively non-polar sorbate, 2,4-decadienal is retained longer on XAD-7 then on the XAD-2 despite the polarity difference in the two crosslinked polymers. This suggests that XAD-4, another crosslinked styrene/divinylbenzene resin having a specific surface area of 725 m^2/g, may be the most preferred synthetic trapping sorbent.

Table II. Adsorbate BTV/Gram on Selected Adsorbent Resins

Conditions: SCF-CO_2, 170 atm., 80°C

Adsorbate	Tenax	XAD-2	XAD-7	XAD-8
Ethyl Caproate	2.56*	6.27	7.52	3.49
Ethyl Caprylate	3.08	7.46	7.72	4.49
2-Pentylfuran	3.01	5.78	6.27	3.99
2,4-Decadienal	5.34	15.0	22.9	12.0
2-Heptanone	3.42	3.99	6.4	3.49
Heptanal	3.15	4.48	4.4	2.76

*All figures in ml of CO_2 at 170 atm., 80°C/gram of resin

The above breakthrough volume data, including values for sorbates minimally retained on activated carbon, can be used to estimate the time for breakthrough to occur on a process sorbent bed. For the gas recycle extractor described earlier, the number of extractions, N, which can be performed without requiring a change in the activated carbon is:

$$N = \frac{(BTV/m)(m)}{(F)(ET)} \quad (3)$$

where m = mass of the adsorbent in grams

ET = extraction time

As an example of the use of Equation 3, the BTV/gram for a limiting case, the highly volatile 2-pentylfuran on activated carbon, may be used to compute the total breakthrough volume for a 650 gram charge of carbon in the recycle stream. The specific breakthrough volume for 2-pentylfuran on activated carbon is 357 ml/gram, quite high compared to the corresponding values listed for the sorbents listed in Table II. This large breakthrough volume is reflective of the affinity exhibited by the activated carbon even in the presence of the supercritical fluid, however it is probably typical of a "light" component which would breakthrough rapidly compared to the other similar but less non-volatile odoriferous species. Division of the total breakthrough volume (232 liters for 2-pentylfuran) by the gas flow rate of 0.443 liters/minute (measured on-line by a mass flow meter) yields a breakthrough time of 8.73 hours. For a typical extraction run, using carbon dioxide at 0.5 lb/min for 20 minutes, this would allow over 26 extractions to be performed before the adsorbent would have to be replaced or regenerated. Using the above CO_2 flow rate and extraction time, 99% of the theoretical oil yield can be realized when extracting flaked soybean seed.

It is interesting to compare the above breakthrough volume result on activated carbon with similar data generated on one of the resinous adsorbents, XAD-7. As shown in Table III, for both a highly retained component, 2,4-decadienal, and 2-pentylfuran; the breakthrough volumes are one to two orders of magnitude smaller than for 2-pentylfuran on activated carbon using the same process conditions and sorbent charge given in the previous example. Therefore, using XAD-7 in place of activated carbon, would allow one to capture 2,4-decadienal, but not 2-pentylfuran in <u>one 20 minute extraction</u>. Clearly, for a process situation where many conditions are fixed, increasing the charge of XAD-7 may be the only convenient way of retaining the 2-pentylfuran for multiple extraction cycles. From an analytical perspective, adjustments would have to be made in either the sampling flow rate, adsorbent weight, or sampling time to trap a representative sample of the fluid phase constituents. The relatively low specific breakthrough volumes of many sorbates on synthetic resin sorbents illustrates the need for investigating alternative sorbents, particularly those which are less expensive and intermediate between XAD-7 and activated carbon in retentive capacity.

Table III. Comparative BTV Results for Different Sorbates on XAD-7 Resin

	2,4-Decadienal	2-Pentylfuran
Specific BTV of sorbate	22.9 ml/g	6.27 ml/g
BTV on XAD-7 column	14.9 L	4.08 L
Breakthrough time on resin column	33.6 min	9.41 min

Conditions: SC-CO_2 Flow Rate - 0.443 L/min, 170 atm, 80°C, Resin wt. = 650 g

Conclusions

An array of factors, such as pressure, temperature, and sorbent type determine the breakthrough volume behavior of adsorbates on adsorbent columns used in process engineering and analytical chemistry. These factors contribute to the complexity of the retention process in the presence of a supercritical fluid mobile phase to such an extent that experimental measurement seems to be the surest way of assessing the required data for particular sorbate/sorbent combinations. The pulse chromatographic method described in this publication offers a rapid and convenient technique for determining BTV data, particularly under the pressure and temperature conditions corresponding to the actual processing conditions. In contrast to chromatographic elution pulse methods conducted at near atmospheric conditions (46), the above technique requires only an extrapolation of the retention volume to process flow conditions.

The breakthrough volume trends for many sorbate types on the porous polymeric sorbents indicate a limited trapping capacity in the supercritical fluid CO_2 above 200 atmospheres. Fractionation and selective retention on these sorbents seems only possible below this specified pressure limit for the odoriferous solutes examined in this study. Adsorbent surface area appears to be the most significant factor contributing to the retention of sorbates on these sorbents as well as activated carbon. For certain synthetic adsorbents (Tenax, XAD-2) employed in this study, pressure-induced morphological changes in the polymer matrix lead to an increase in the sorption capacity, and hence to an increase in breakthrough volumes at intermediate pressures.

The breakthrough volume data base generated in this study has application in both process engineering and analytical chemistry. For example, it appears that the polymeric sorbents examined in this work are unsuitable for long term processing applications requiring elevated pressures, however by employing the conditions and constraints defined in this study, such sorbents could be used to characterize the constituents in a supercritical fluid process stream using analytical sorbent trapping techniques. The larger breakthrough volumes recorded for specific sorbates on activated carbon compared to the polymeric media suggests that a longer service lifetime would be realized by employing the carbonaceous sorbent. However, certain other features of activated carbon/sorbate interactions, such as irreversible adsorption at lower pressures, lengthy desorption kinetics and chemical contamination limit the analytical use of this sorbent.

Acknowledgment
The expertise and assistance of F. L. Baker in the SEM studies is gratefully acknowledged.

The mention of firm names or trade products does not imply that they are endorsed or recommended by the U.S. Department of Agriculture over other firms or similar products not mentioned.

Literature Cited

1. Zosel, K. In *Extraction with Supercritical Gases*; Schneider, G. M.; Stahl, E.; Wilke, G., Eds.; Verlag Chemie: Deerfield Beach, Florida, 1980; pp 20-22.
2. Shishikura, A.; Fujimoto, K.; Kaneda, T.; Arai, K. *Agric. Biol. Chem.* 1968, *50*, 1209.
3. King, J. W. In *Supercritical Fluids: Chemical and Engineering Principles and Applications*; Squires, T. G.; Paulaitis, M. E., Eds.; American Chemical Society: Washington, DC, 1987; pp 150-171.
4. Modell, M.; deFilippi, R. P.; Krukonis, V. J. In *Activated Carbon Adsorption of Organics from the Aqueous Phase*; Suffet, I. H.; McGuire, M. J., Eds.; Ann Arbor Science Publishers: Ann Arbor, Michigan, 1980; Vol. 1, pp 447-461.
5. deFilippi, R. P.; Krukonis, V. J.; Robey, R. J.; Modell, M. *Supercritical Fluid Regeneration of Activated Carbon for Adsorption of Pesticides*; EPA Report 600/2-80-054: 1980.
6. Hawthorne, S.B.; Miller, B. J. *J. Chromatog. Sci.* 1986, *24*, 258.
7. Wright, B. W.; Wright, C. W.; Gale, R. W.; Smith R. D. *Anal.Chem.* 1987, *59*, 38.
8. Raymer, J. H.; Pellizzari, E. D. *Anal. Chem.* 1987, *59*, 1043.
9. Friedrich, J. P.; List, G. R.; Heakin, A. J. *J. Am. Oil Chemists Soc.* 1982, *59*, 288.
10. Christianson, D. D.; Friedrich, J. P.; List, G. R.; Warner, K.; Bagley, E. B.; Stringfellow, A. C.; Inglett, G. E. *J. Food Sci.* 1984, *49*, 229.
11. Gallant, R. F.; King, J. W.; Levins, P. L.; Piecewicz, J. F. *Characterization of Sorbent Resins for Use in Environmental Sampling*; EPA Report 600/7-78-054: 1978.
12. Novak, J.; Visak, V.; Janak, J. *Anal. Chem.* 1965, *37*, 660.

13. Melcher, R. G.; Peters, T. L.; Emmel, H. W. In **Analytical Problems**; Springer-Verlag: New York, NY, 1986; pp 82-84.
14. Vidal-Madjar, C.; Gennord, M; Benchah, F.; Guiochon, G. **J. Chromatog. Sci.** 1978, 16, 190.
15. Butler, L. D.; Miller, D. J. **J. Chromatog. Sci.** 1976, 14, 117.
16. Ballou, E. V. **Second NIOSH Solid Sorbents Roundtable**; HEW Report 76-193, 1976.
17. Simon, G. G.; Bidleman, T. F. **Anal. Chem.** 1979, 51, 1110.
18. Reineccius, G. A. In **Flavor Chemistry of Fats and Oils**; Min, D. B.; Smouse, T. H., Eds.; American Oil Chemists Society: Champaign, Illinois, 1985; p. 268.
19. DuPay, H. P.; Flick, G. J.; Bailey, M. E.; St. Angelo, A. J.; Legendre, M. G.; Sumrell, G. **J. Am. Oil Chemists Soc.** 1985, 62, 690.
20. Aria, S.; Noguchi, M.; Yamashita, M.; Kato, H.; Fujimaki, M. **Agric. Biol. Chem.** 1970, 34, 1569.
21. Frankel, E. N. In **Flavor Chemistry of Fats and Oils**; Min, D. B.; Smouse, T. H., Eds.; American Oil Chemists Society, Champaign: Illinois, 1985; pp. 1-37.
22. Aspelund, T. G.; Wilson, L. A. **J. Agric. Food Chem.** 1983, 31, 53.
23. McHugh, M. A.; Krukonis, V. J. **Supercritical Fluid Extraction: Principles and Practice**; Butterworths: Boston, MA, 1986; pp. 237-239.
24. Conder, J. R.; Young, C. L. **Physicochemical Measurements by Gas Chromatography**; John Wiley & Sons: New York, NY, 1979; pp. 26-31.
25. Ecknig, W.; Polster, H. J. **Separation Sci.** 1986, 21, 139.
26. Sie, S. T.; Van Beersum, W.; Rijnders, G. W. A. **Separation Sci.** 1966, 1, 459.
27. Groninger, G.; Hedden, K.; Rao, B. R. **Chem. Eng. Tech.** 1987, 10, 63.
28. Reucroft, P. J.; Sethuraman, A. R. **Energy & Fuels.** 1987, 1, 72.
29. Ender, D. H. **Chem. Tech.** 1986, 16, 52.
30. Boyer, R. F.; Spencer, R. S. **J. Polym. Sci.** 1948, 397.
31. Liau, I. S.; McHugh, M. A. In **Supercritical Fluid Technology**; Penninger, J. M. L.; Radosz, M.; McHugh, M. A.; Krukonis, V. J., Eds.; Elsevier, NY, 1985, pp. 415-434.
32. Chiou, J. S.; Barlow, J. S.; Paul, D. R. **J. Appl. Polym. Sci.** 1985, 30, 2633.
33. Springston, S. R.; David, P.; Steger, J.; Novotny, M. **Anal. Chem.** 1986, 58, 997.
34. Novotny, M.; David, P. **J. High Res. Chromatog. Chromatog. Commun.** 1986, 9, 647.
35. Wang, W. V.; Kramer, E. J.; Sachse, W. H. **J. Polym. Sci., Polym. Sci. Ed.** 1982, 20, 1371.
36. Hojo, H.; Findley, W. N. **Polym. Eng. Sci.** 1973, 13, 255.
37. Sefcik, M. D. **J. Polym. Sci., Polym. Phys. Ed.** 1986, 24, 957.

38. Albright, R. L.; Jakovac, I. J. Catalysis by Functionalized Porous Organic Polymers; Bulletin IE-287; Rohm & Haas Company: Philadelphia, PA, 1985, pp. 4-6.
39. McHugh, M. A.; Krukonis, V. J. Supercriticl Fluid Extraction: Principles and Practice; Butterworths: Boston, MA, 1986, pp. 230-235.
40. Chester, T. L.; Innis, D. P. J. High Res. Chromatog. Chromatog. Commun. 1985, 8, 561.
41. Sanetra, R.; Kolarz, B. N.; Wiochowicz, A. Polymer. 1985, 26, 1181.
42. Sanetra, R.; Kolarz, B. N.; Wiochowicz, A. Angew. Makromol. Chem. 1986, 140, 41.
43. Tiltscher, H.; Wolf, H.; Schelchshorn, J. Angew. Chem. Int. Ed. Engl. 1981, 20, 892.
44. Mattson, J. S.; Mark, H. B. J. Colloid Interface Sci. 1969, 31, 131.
45. Ananymous Making Waves; IWT Company: Rockford, IL, 1984, 2, 1.
46. Adams, J.; Menzies, K. T.; Levins, P. Selection and Evaluation of Sorbent Resins for the Collection of Organic Compounds; EPA Report 600/7-77-044: 1977.

RECEIVED December 21, 1987

Chapter 5

Concentration of Omega-3 Fatty Acids from Fish Oil Using Supercritical Carbon Dioxide

S. S. H. Rizvi, R. R. Chao, and Y. J. Liaw

Institute of Food Science, Cornell University, Ithaca, NY 14853

> Recent studies on the role of omega-3 fatty acids in fitness and health have stimulated considerable interest in the development of supercritical fluid extraction processes for concentrating them from marine oils. The basic approaches utilized by several researchers in their attempts to realize this goal are reviewed in this paper. The various parameters influencing the purity and yield of eicosapentaenoic and docosahexaenoic acids obtained from fish oils with different pretreatments are discussed.

The total world-wide production of fish oil in 1985 was reported at 1.4 million metric tons (1). Fish oil is utilized mainly in food and pharmaceutical formulations. Less than five percent is used for such diverse applications as the production of paints, glues, preservatives, lubricants, cosmetics or as an energy source. In the U.S., about 30-40% of the catch is converted into fish meal and oil and, according to the most recent figures, the 1985 fish oil production totaled about 129 thousand metric tons with 98% contributed by menhaden (2). Apart from a few therapeutic products, such as cod liver oil, fish oil is not approved for human consumption in the United States. As a result, over 95 percent of U.S. fish oil is exported overseas where it is used in margarine and other foods. However, a petition is currently before the Food and Drug Administration to have menhaden and hydrogenated menhaden approved as GRAS.

The position of fish oil in the market is affected by certain specific factors, some of which apply to most oils while others are peculiar to fish oil alone. First, apart from its use for pharmaceutical purposes, fish oil, because of its high content of polyenic acids, must be hydrogenated before use in food formulations. Second, legislation meant mainly to prevent usage of erucic acid may in some countries encompass all C22:1 acids and thus affects the usage of fish oils containing cetoleic acid. Third, the quality of crude fish oil may be more variable in terms of free fatty acids, color, odor, etc. than that of other oils and may pose more problems for the refiner and hydrogenator. These factors tend to restrict the use of fish oil and have, therefore, generally reduced its value.

Recently, fish oils have attracted wide commercial and academic interests as a rich source of polyunsaturated fatty acids, particularly C20:5 ω-3 (eicosapentaenoic acid, EPA) and C22:6 ω-3 (docosahexaenoic acid, DHA), which are reported to possess potential therapeutic advantages (3-7). While public awareness of the nutritional value of fish oil may increase seafood consumption, it is also anticipated that a market will develop for marine oils for direct use in diets. For example, fish oil capsule sales are predicted to increase two to five times over the next few years to reach an ultimate $500-million-a-year mark (8). In a recent study (9) on the dose-response relationship between omega-3 fatty acids intake and certain blood parameters in human subjects, omega-3 fatty acid concentrate was preferred to "whole" fish oil because the former keeps the daily intake of total fatty acids lower thus minimizing the ingestion of physiologically undesirable fatty acids.

All of these studies have stimulated considerable interest in the development of efficient methods for concentrating omega-3 polyunsaturated fatty acids from marine oils. One promising, state-of-the-art process is supercritical fluid extraction (SFE), the topic of this symposium. In addition to its potential for cleaning and purification of fish oil, SFE also offers attractive possibilities to selectively concentrate desirable omega-3 fatty acids. This paper is primarily aimed at reviewing the current status of fish oil fractionation using the SFE technique.

Families of Unsaturated Fatty Acids and Fish Oil Composition

The principal families of unsaturated fatty acids are shown in Figure 1. Those important in fish oil are the omega-3 fatty acids: C18:4, C20:4, C20:5, C22:4, C22:5, and C22:6. By definition, omega-3 or n-3 means that the first double bond begins at the third carbon from the methyl end of the chain. In the "number:number" designation, the first number designates chain length and the second number designates how many double bonds are present.

The omega-3 fatty acid composition of selected fish oils is shown in Table I. The various species ranging from lean to fatty fish contain from 0.7 to 15.5% oil. Omega-3 fatty acids generally account for 25 to 30% of the total lipid content in most fish (11). Certain major fatty acids vary widely among the species, e.g., 1.6-8.0% myristic acid; 0.5-33.4% palmitic acid; 2.0-11.2% palmitoleic acid; 5.2-29.1% oleic acid; 0.7-10.5% eicosenoic acid;

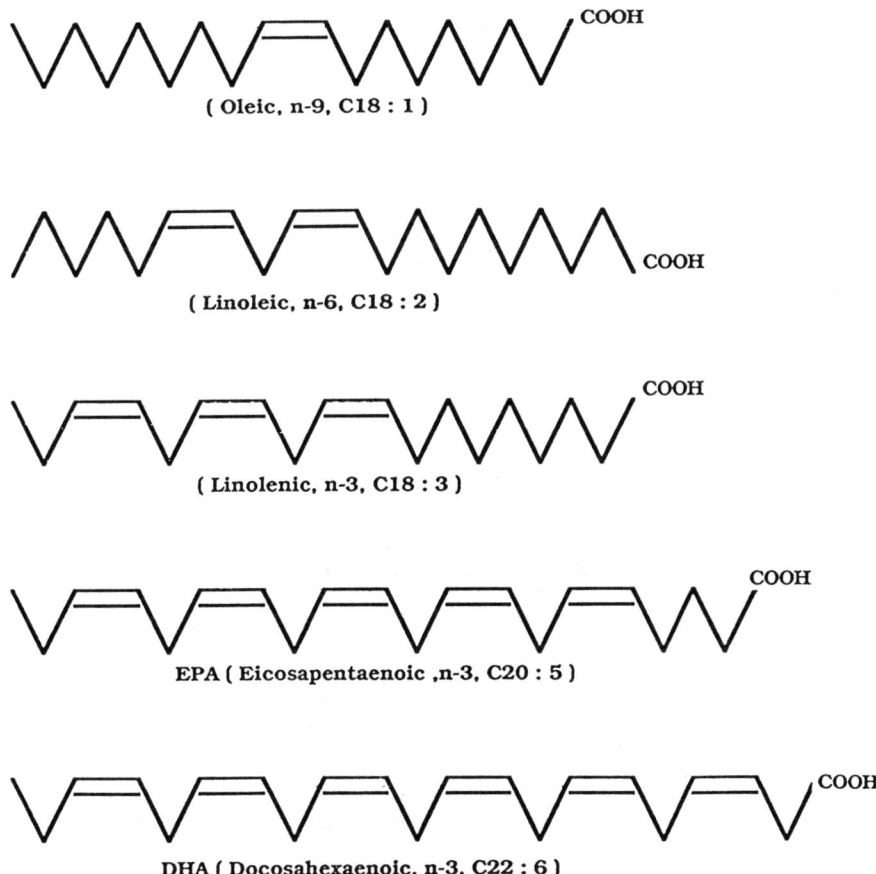

Figure 1. Families of unsaturated fatty acids

0.2-11.6% docasenoic acid; 5.0-21.5% eicosapentaenoic acid and 5.9-26.2% docosahexaenoic acid (12).

Table I. Omega-3 Fatty Acid Composition (%) of Selected Fish Oil (Adapted from Ref. 9.)

Omega-3 Fatty Acids	Herring North sea	Anchovy Peru	Sardine Portugal	Menhaden U.S.A.
16:3	-	-	0.20	0.20
18:3	2.00	0.75	1.00	1.30
18:4	3.15	3.05	3.15	2.75
20:4	0.75	0.70	1.05	1.35
20:5	7.45	17.00	11.00	11.50
22:3	-	0.15	0.15	0.15
22:4	0.25	0.55	0.70	0.50
22:5	0.75	1.60	1.30	1.90
22:6	6.75	8.75	13.00	9.10
Total Omega-3	21.10	32.55	31.55	28.75

Current Methods of Fractionating Fish Oil

On hydrolysis, fish oils yield a mixture of fatty acids derived mainly from mixed glycerides. Separation of these fatty acids based on molecular weight or degree of unsaturation is complicated by several factors. First, the relatively small differences in their molecular weights make it difficult to separate them by conventional means, particularly when saturated and unsaturated fatty acids of the same chain length are to be separated. Second, polyunsaturated compounds are readily susceptible to polymerization, degradation and/or oxidation, even at moderately elevated temperatures.

Current methods for fractionating fish oil include selective removal of saturated as well as mono-unsaturated fatty acids such as C20:1 and C22:1 by urea complexing, adsorption, chromatography, and fractional and/or molecular distillation processes. These are cumbersome and time consuming. Particularly undesirable are methods which require the use of difficult-to-remove organic solvents from the finished products. Use of high temperatures also introduces the possibility of alteration of the fatty acids and formation of toxic derivatives.

Supercritical Fluid Extraction of Fish Oil

Although various supercritical fluids have been found useful as solvents for fatty acids and their esters, carbon dioxide is thus far the most commonly used extractant because of its inherent advantages. Extraction with carbon dioxide is effective at moderately low temperatures, which limits autoxidation, decomposition and polymerization of the highly unsaturated fatty acids present in fish oils. Furthermore, the inert atmosphere of carbon dioxide inhibits autoxidation. Other specific advantages have been extensively described and discussed elsewhere (13-15).

In general, the SFE techniques used to date for fish oil fractionation can be grouped into the following categories, based on operational characteristics:

Single Pass System. The single pass system for fractionation of fish oil operates at a given temperature and utilizes a stepwise adjustment of pressure in the extraction vessel and a lower pressure in the collector. A schematic of the process is shown in Figure 2. A stepwise increase in the pressure of the extraction vessel enhances dissolution of the material to be extracted into the supercritical phase in the order of increasing boiling point or molecular weight. The dissolved material is then separated either isobarically or isothermally in the collector.

Using such a system for separation of three types of fish oils (menhaden, herring and anchovy), Krukonis (Krukonis, V.J. Paper presented at the 75th Annual Meeting of the American Oil Chemist's Society, Dallas, April 29-May 3, 1984) reported the recovery of 97.1% of the C20 and 93.6% of the C22 fatty acids when the starting material was the methyl esters of fish oil; 18.9% C20 and 24% C22 when the fish oil fatty acids were used as the feed stock; and 85.6% C20 and 94.1% C22 when the fish oil triglycerides were used. However, the purity of EPA and DHA was less than desirable, and the concentrations of EPA and DHA were not enhanced to any significant degree. The above work showed that fatty acids were most conveniently separated as their methyl esters. The solubility of fish oil and the fatty acid esters increased as the supercritical carbon dioxide pressure was increased (Figure 3). It is evident from Figure 3 that SFE at higher temperatures led to a lower solubility of the fish oil until pressure of about 5500 psi (37.9 MPa) was reached, beyond which solubility in the supercritical phase increased. It is also apparent from Figure 3 that methyl esters of fish oil because of their higher vapor pressure show higher solubility than fish oil at any given operating temperature and pressure condition.

Refluxing Systems - Variable Temperature. Figure 4 illustrates a scheme for supercritical fluid extraction involving the use of extraction and separation vessels with an intermediate heat exchanger, also known as a "hot finger", and a fractionation column (16). The selectivity of the fractionation of the various fatty acids is enhanced by the hot finger which heats the supercritical carbon dioxide thus reducing its density. As a result, the solubility of all solutes in the supercritical fluid is decreased, but not to the same extent. The less soluble components return to the extraction vessel while the more soluble components pass into the separation vessel and are recovered by precipitation at a reduced pressure. As shown in Table II, fractions 2-7 obtained in the first step contained mostly C18 and C20 esters, and fractions 9-11 contained C20 and C22 esters. These fractions were then combined to give the starting material for the second fractionation step. The first and last fractions of the second step which contained only the C14 to C18 and the C22

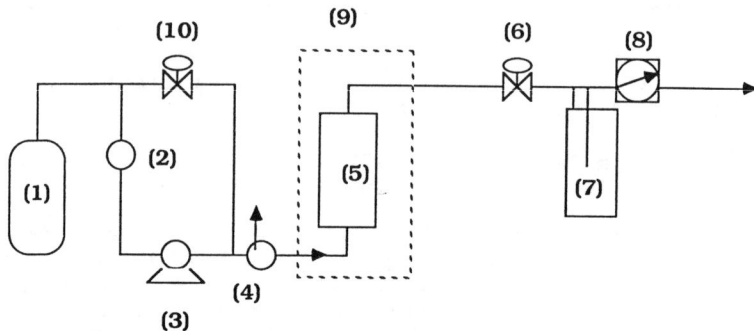

1. Gas Cylinder
2. Condenser
3. Plunger Pump
4. Heat Exchanger
5. Extraction Vessel
6. Metering Valve
7. Collector
8. Dry Gas Meter
9. Water Bath
10. Back Pressure Regulator

Figure 2. Schematic of a single pass supercritical fluid extraction system

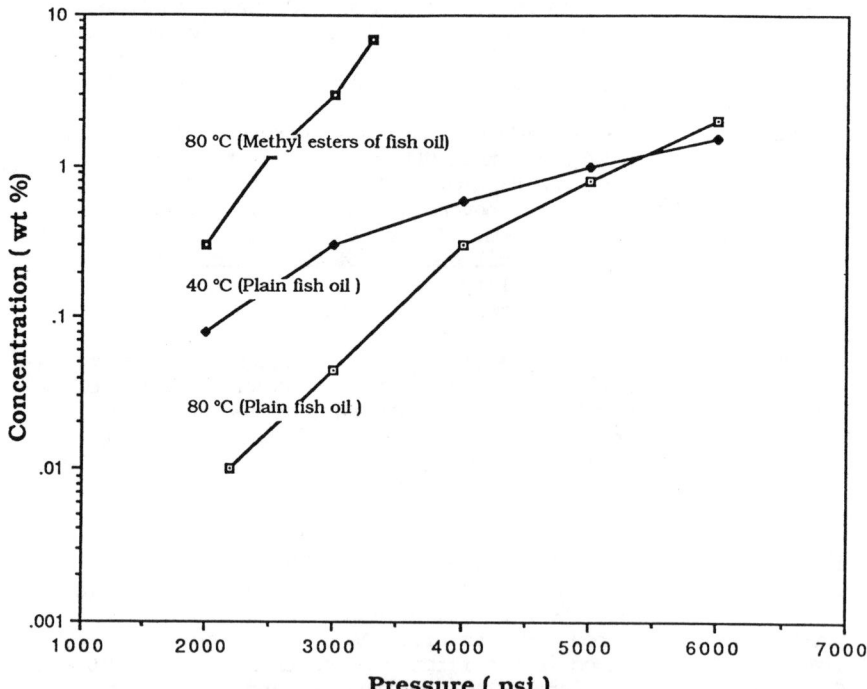

Figure 3. Solubility of fish oil in supercritical CO_2. (Reproduced with permission from Ref. 19. Copyright 1984 American Oil Chemists' Society.)

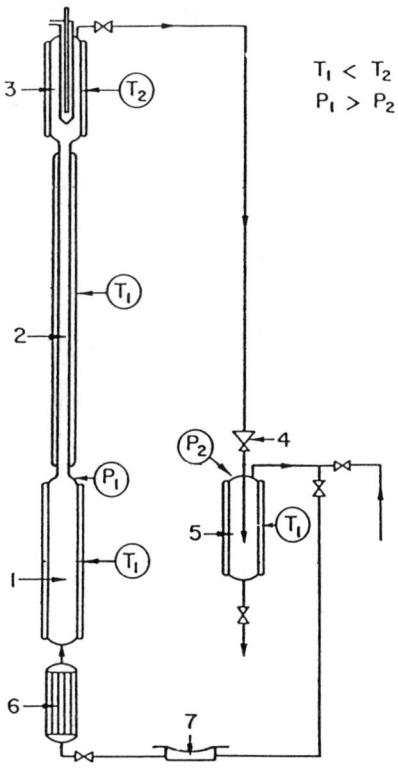

Extraction Vessel (1), Column (2), Hot Finger(3),
Expansion Valve (4), Separation Vessel (5),
Heat Exchanger (6), Membrane compressor (7).

Figure 4. Schematic of a supercritical fluid extraction unit with hot finger (Reproduced with permission from Ref. 16. Copyright 1984 Verlag Chemie.)

esters, respectively. The purities of C20 and C22 esters of the combined fraction 8 from the first step with fraction 3 from the second step were 96.2% and 93.6%, respectively.

Table II. Composition of Fractions Obtained from Cod Liver Oil by the Two-step SFE Technique

First Step Fractions	wt.(g)	%C14	%C16	%C18	%C20	%C22
1	911.6	11.0	48.9	37.2	–	–
2	122.6	–	3.6	92.0	2.2	–
3	97.6	–	1.9	70.1	26.0	–
4	68.7	–	–	46.2	52.8	–
5	55.5	–	–	34.2	64.7	–
6	54.1	–	–	23.6	75.5	–
7	70.0	–	–	13.3	85.3	–
8	305.8	–	–	2.6	95.8	0.7
9	41.2	–	–	–	72.6	25.2
10	73.0	–	–	–	35.8	89.8
11	127.1	–	–	–	9.2	89.8
12	70.0	–	–	–	–	84.5
Second step Fractions						
1	146.8	–	1.3	76.1	20.8	–
2	99.4	–	–	25.3	73.3	–
3	122.0	–	–	1.6	97.5	0.6
4	62.0	–	–	–	48.6	51.3
5	139.8	–	–	–	0.6	98.0

Based on initial C20 charge:
  Total Yield: 67.7%
  Purity   : 96.2%

Source: Reproduced with permission from Ref. 16. Copyright 1984 Verlag Chemie.

Refluxing System - Variable Temperature and Pressure. This approach, shown in Figure 5, allows for regulation of both temperature and pressure during refluxing (Daniels, J.A.; Rizvi, S.S.H.; Black, J.M., and German, J.B. Cornell University, Ithaca, N.Y. 1986). In the reflux loop, the temperature and/or pressure can be varied to reduce the solubility of solutes so they form a condensed liquid fraction which is, in turn, pumped back to the top of the rectification column to establish a counter-current flow. The supercritical solvent containing the remaining solute is routed to the separation vessels where the extract is collected under reduced pressure and temperature conditions. Solvent flow rate and volume are measured by a rotameter and flow totalizer downstream from the separation vessel. Starting with free fatty acids of herring fish oil, Daniels et al. showed (Table III) that omega-3 free fatty acid concentration was increased from 40.4% to 87.8%. The condition of supercritical carbon dioxide was 1,400 psi(9.65 MPa) and 35°C in the extractor and 1,350 psi (9.3 MPa) and 95°C in the reflux loop. However, the concentrations of EPA and DHA increased to only 22.5% from 17.32% and to 29.23% from 9.49% respectively. There are two reasons for using free fatty acids as the starting material. First, the process would be simpler and less costly without the methylation step. Second, free fatty acids are more compatible with possible food related applications than are methyl esters.

Table III. Fatty Acid Composition, and Omega-3 Mass Balance for the Original Fatty Acid Sample, Extract Obtained with Refluxing, and Concentrate of Omega-3 Product

	Original sample wt(%) and wt(g)		Extract wt(%) and wt(g)		Concentrate wt(%) and wt(g)		% Recovery
wt(g)	100	10.0	61	6.1	30	3.0	91.0
Fatty Acids							
C18:3	5.86	0.589	1.82	0.110	11.81	0.354	60.4
C20:3	4.48	0.403	0.38	0.023	14.23	0.354	95.3
C20:4	1.03	0.103	0.02	0.001	2.10	0.063	61.2
C20:5	17.30	1.730	1.08	0.066	22.25	0.667	38.6
C22:5	2.20	0.220	0.01	-	8.21	0.220	100.0
C22:6	9.49	0.949	0.43	0.026	29.23	0.877	92.4
Total	40.40	4.030	3.74	0.277	87.84	2.590	64.3

Source: Reproduced with permission from Ref. 21. Copyright 1986 Cornell University.

Comparison of the Yield with SFE of Omega-3 Fatty Acids from Various Starting Materials

Comparing the results of Krukonis and Daniels et al. shows that when the starting material was free fatty acids, the recovery of EPA was similar but low. For DHA, the recovery reported by Krukonis was very low, around 19%, and significantly different from that of Daniels et al. at 92.4%. Comparison of the results of Krukonis to those of Eisenbach (16) for the fractionation of methyl esters indicates that both EPA and DHA show very similar concentrations (Table IV). From Table IV it is apparent that SFE of fish oil enhances EPA and DHA concentrations and that fatty acids are more efficiently separated as their methyl esters.

Table IV. Comparison of Yields of Omega-3 Fatty Acids from Various Starting Materials

Starting material	Yield (%)	Recovery EPA(%)	DHA(%)
Triglycerides	85.7	63.6	45.8
Triglycerides (Extraction followed by immediate analysis)	96	94.1	85.6
Free fatty acids	97.7 (64.3)*	24 (38.6)*	18.9 (92.4)*
Methyl esters	>100 (67.7)**	97.1 (96.2)**	93.6 (93.6)**

(From Krukonis, V.J. Paper presented at the 75th Annual Meeting of the American Oil Chemist's Society, Dallas, April 29-May 3, 1984)
*(From Daniels, J.A.; Rizvi, S.S.H.; Black, J.M.; German, J.B. Cornell University, Ithaca, NY, 1986)
**(From Ref. 16.)

SFE of Urea Preconcentrated Samples

Since fish oil contains a broad mixture of saturated and unsaturated fatty acids, including abundant amounts of shorter chain, saturated fatty acids, a preliminary separation step prior to SFE may be an attractive proposition. And, since supercritical carbon dioxide fractionates fatty acids mainly on the basis of molecular weight; i.e., it distinguishes more readily between C18:1, C20:1, and C22:1 than it does between C20:0, C20:1, C20:4, and C20:5 fatty acids, the choice of starting material affects the degree of concentration obtained. In order to efficiently concentrate the desired polyunsaturated fatty acids, the starting fish oil should be as free as possible from interfering fatty acids with the same number of carbons. To increase the concentration of EPA, it is possible to pretreat the esters and/or fatty acids of fish oil with urea and methanol to remove saturated, mono- and diunsaturated components prior to SFE.

Urea crystallizes in a compact tetragonal pattern, which may form inclusion compounds with aliphatic normal chain substances (17). Preconcentration of the fatty acids with urea is achieved on the basis of degree of unsaturation; the more unsaturated the fatty acid, the less it will be included in the urea crystal. The equilibrium reaction of the urea complex is shown in Figure 6. The "induced" hexagonal structure is stable only in the presence of the included compound. Removal of the included partner by extraction or volatilization causes a rapid breakdown of the urea lattice. Aliphatic hydrocarbons, alcohols, ketones, esters, ethers, amines, nitriles, mono- and dicarboxylic acids, and their halogen derivatives can be included in the urea complex.

5. RIZVI ET AL. *Concentration of Omega-3 Fatty Acids*

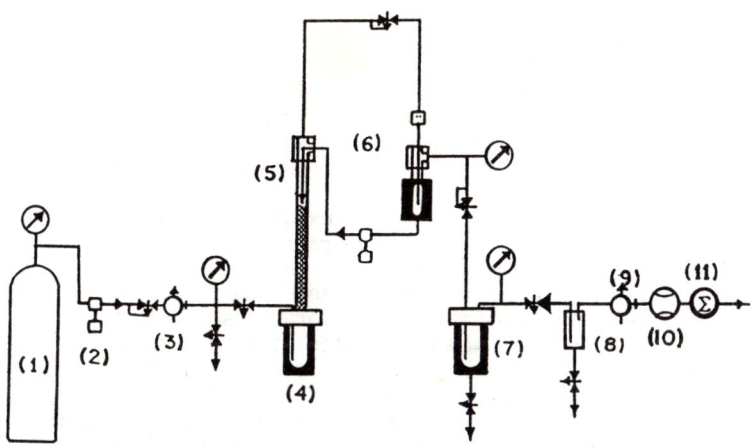

CO2 Supply(1), Feed Pump(2), Heat Exchanger(3), Extraction Vessel(4), Rectification Column(5), Reflux Loop(6), Separation Vessel(7 & 8), Cold Trap(9), Rotometer(10), and Flow Totalizer(11).

Figure 5. Schematic of a supercritical fluid extraction system with reflux. (Reproduced with permission from Ref. 21. Copyright 1986 Cornell University.)

(m) Urea + (n) Guest Molecule ⇌ Inclusion Compound
   (Fatty Acid)   (Solid)

Figure 6. Equilibrium reaction of urea

Unsaturation of the chain does not inhibit adduct formation but lessens the stability of the complex.

In the fatty acid series, chain length (molecular weight) and unsaturation are opposite with respect to complex stability, i.e., shorter chain lengths and a greater number of double bonds lead to less complex stability. Trans isomers form more stable adducts than the corresponding cis isomers and compounds with conjugated double bonds complex better than those with isolated double bond.

The composition of urea complexes is usually represented by the ratio of moles of urea per mole of included compound. The ratio is constant for any given complex and is independent of the relative concentrations of the partners previous to adduct formation and of the temperature; that is, urea adducts must be considered as true compounds and not as mixed crystals. Domart (17) indicated that there was a considerable selectivity in the fatty acids precipitated when the moles of urea present were less than the quantity required for maximum precipitation. For example, if the mole ratio of urea to fatty acids was 4.6, then the fatty acids obtained from the complex are highly saturated. However, at higher mole ratios, not only saturated fatty acids, but also some portion of the unsaturated fatty acids present were complexed by urea. Virtually all the saturated and monoenoic fatty acids are precipitated at a mole ratio in the region of 12:1 to 13:1 (Table V).

Table V. Fractionation of Menhaden Oil with Different Mole Ratios of Urea

Mole ratio urea to fatty acid	% Yield (fatty acids in complexes)	%Yield (fatty acids in filtrate)
4.6 : 1	11.6	80.8
9.1 : 1	29.6	61.6
13.8 : 1	49.4	41.6
18.4 : 1	61.0	36.4
23.0 : 1	63.0	34.2

(Adapted from Ref. 17.)

Adduct formation is exothermic; the heat of reaction and the stability of the complex increase with chain length and straightness. Thus, the reaction is displaced toward dissociation of the complex when the temperature is raised. High concentrations of urea are necessary to form the adducts. In practice, only saturated urea solutions are used, because an excess of the solvent (methanol) used easily inverts the reaction. However, complex formation is never complete. Even with pure substances, a certain residual concentration of the included compound is stable in a saturated urea solution. This limitation should be born in mind for most of the analytical applications of the adducts. In dealing with a mixture of adduct-forming substances, the differences in the values of reaction constants from substance to substance may be great enough to allow for selective crystallization. The reactions are competitive and an equilibrium is reached for the complexes according to their stability. However, in precipitating the complex, some of the excess components of the mixture, including those substances which do not generally form adducts, are entrained or adsorbed.

Nilsson et al. (Nilsson, W.; Hudson, J.K.; Stout, J.S., and Gauglitz, E.J. Paper presented at the 77th Annual Meeting of the American Oil Chemist's Society, Honolulu, May 19, 1986) demonstrated that using a temperature gradient along the SFE column gave a better separation of urea-concentrated ethyl esters of menhaden fish oil fatty acids than using a constant temperature (Figure 7). In Figure 7, a temperature gradient is established such that $T_4 > T_3 \geq T_2 \geq T_1$. Carbon dioxide enters the column and flows upwards through the gradient. Under these conditions the ester solubility decreases with increasing temperature, resulting in reflux. Although Nilsson et al. did not mention the urea concentration used, the ratios of urea and methanol to the organic substance ought to have been high in order to improve the purity or the yield of adduct. Generally, three to six parts of urea and seven to twenty parts of methanol are used for every part of the sample (17). Increasing the ratio of urea to substance improves the yield of adduct, but the resulting precipitate is contaminated with urea.

Comparison of Yield of Omega-3 Fatty Acids with and without Urea Pretreatment

Short chain fatty acids of fish oil are mostly saturated. C18 fatty acids are composed primarily of saturates and monoenes. On the other hand, C20 esters are mostly unsaturated. Figure 8 shows fractionation curves by carbon number of the fatty acids with and without urea pretreatment obtained on a SFE system with a temperature controlled column. The substantial overlap between the C18 and C20 curves, illustrates the difficulty of obtaining C20 esters in better purity. Urea complexing removes these compounds preferentially and minimizes the problem of C18-C20 overlap. The C20 curve is sharpened relative to C18 curve giving a fraction of EPA with greater purity.

Supercritical Fluid Extraction with a Clathrate Vessel

The procedure of forming inclusion compounds requires several urea precipitation steps along with filtration, washing, and recovery of the fraction enriched in unsaturated fatty acids. These

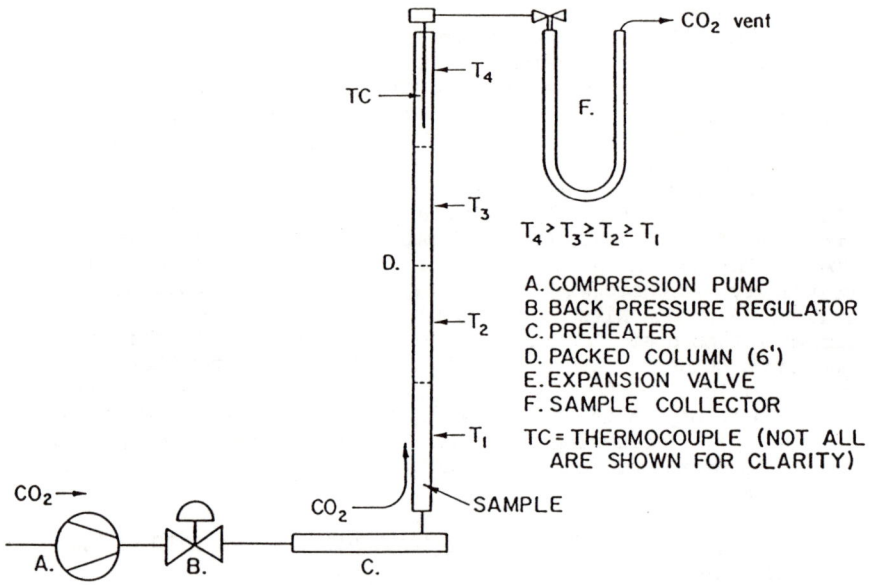

Figure 7. Supercritical fluid extraction with temperature gradient column. (Reproduced with permission from Ref. 20. Copyright 1986 American Oil Chemists' Society.)

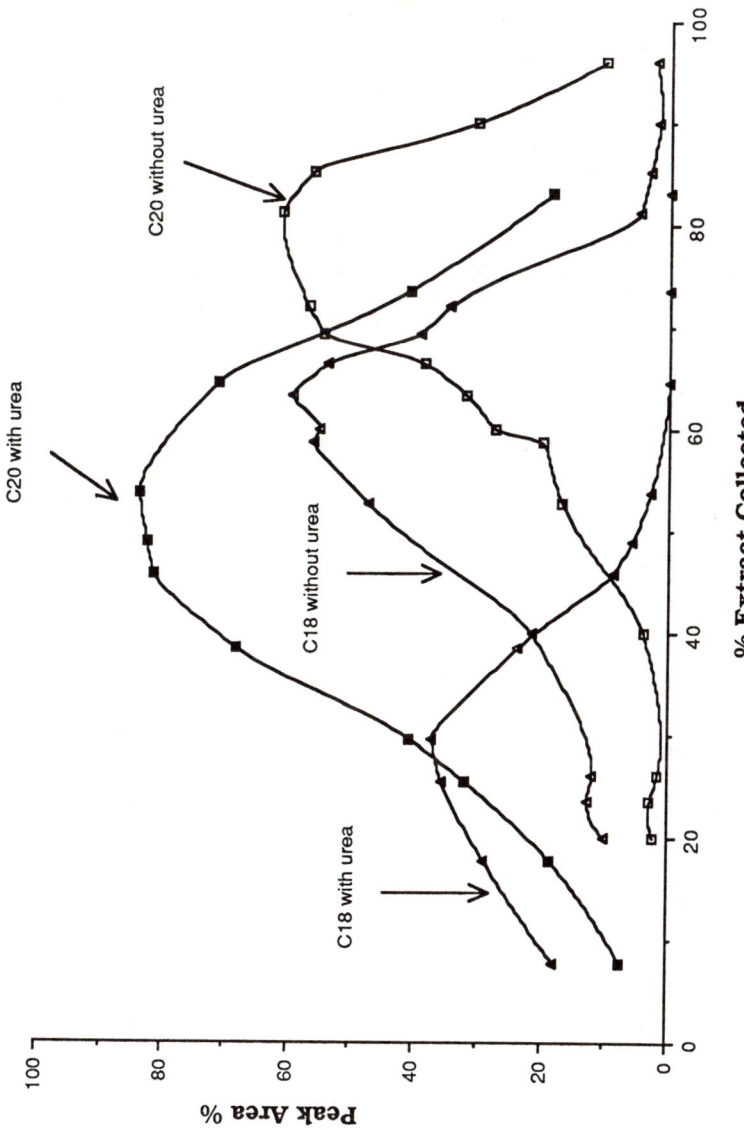

Figure 8. Comparison of yield of omega-3 fatty acids of menhaden oil with and without urea pretreatment. (Reproduced with permission from Ref. 20. Copyright 1986 American Oil Chemists' Society.)

complex operations can be simplified by using supercritical fluid as the solvent for urea adduct formation (18). A supercritical fluid extraction apparatus with a clathrate vessel is shown in Figure 9. In this set-up, the supercritical fluid with dissolved free fatty acids or methyl esters goes into a vessel containing finely powdered urea. Clathrate compounds are formed and unreacted solutes exit with the supercritical fluid. The fluid is then depressurized through a heated metering valve and the extract is collected in a trap.

Figures 10 and 11 show the effluent composition, mainly C20 and C22 fatty acids methyl esters derived from sardine oil. It can be seen from these figures that the higher unsaturated fatty acid methyl esters are enriched in the fluid phase of the clathrate forming vessel. Urea adducts in supercritical carbon dioxide solvent are formed according to their level of unsaturation similar to that in methanol. According to Saito (18), maximum adduct formation occurs around 40°C in supercritical carbon dioxide.

Conclusions

The feasibility of obtaining fractions of fish oil rich in EPA and DHA by supercritical fluid extraction has been shown. However, a single-pass system is not adequate to enhance the selectivity of the process. Modifications of a SFE unit to include a "hot finger" has been shown to provide a higher degree of selectivity. Polyunsaturated fatty acid esters have been demonstrated to possess a higher solubility in supercritical carbon dioxide. Also described in this paper is a SFE system with both temperature and pressure-induced refluxing capabilities; a process whereby total omega-3 polyunsaturated fatty acid concentrations in excess of 87% were obtained starting with free fatty acids. Use of free fatty acids should be desirable and economical since they are more food compatible than their esters. Preconcentration of fish oil fatty acids or their esters with urea and methanol prior to supercritical fluid extraction has been found to yield better EPA and DHA purity. An alternative process has also been developed where supercritical carbon dioxide saturated with fatty acids or their esters is mixed directly with finely powdered urea. A section of the tower could be operated under an appropriate temperature gradient to decompose the clathrate compounds, strip the unsaturated fatty acids, and reactivate the host material.

Based on current knowledge it is reasonable to expect a continued expansion of demand for fish oils rich in EPA and DHA. Successful commercial realization of this trend for comparative economic advantages to the fish industry may hinge on the ability to use new processing technologies to produce stable, clean, edible fish oil as well as its enriched concentrates. Supercritical fluid extraction techniques along with other adjunct technologies, such as microencapsulation, hold the promise of delivering quantities of high quality fish oil at reasonable costs.

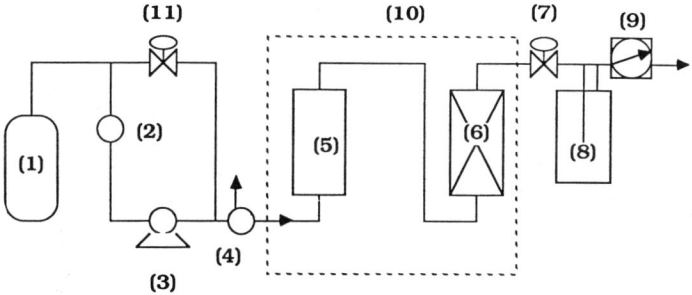

1. Gas Cylinder
2. Condenser
3. Plunger Pump
4. Heat Exchanger
5. Extraction Vessel
6. Clathrate Vessel
7. Metering Valve
8. Collector
9. Dry Gas Meter
10. Water Bath
11. Back Pressure Regulator

Figure 9. Schematic of a supercritical fluid extraction apparatus with clathrate vessel (Reproduced with permission from Ref. 18. Copyright 1986 Gakkai Shuppan Senta.)

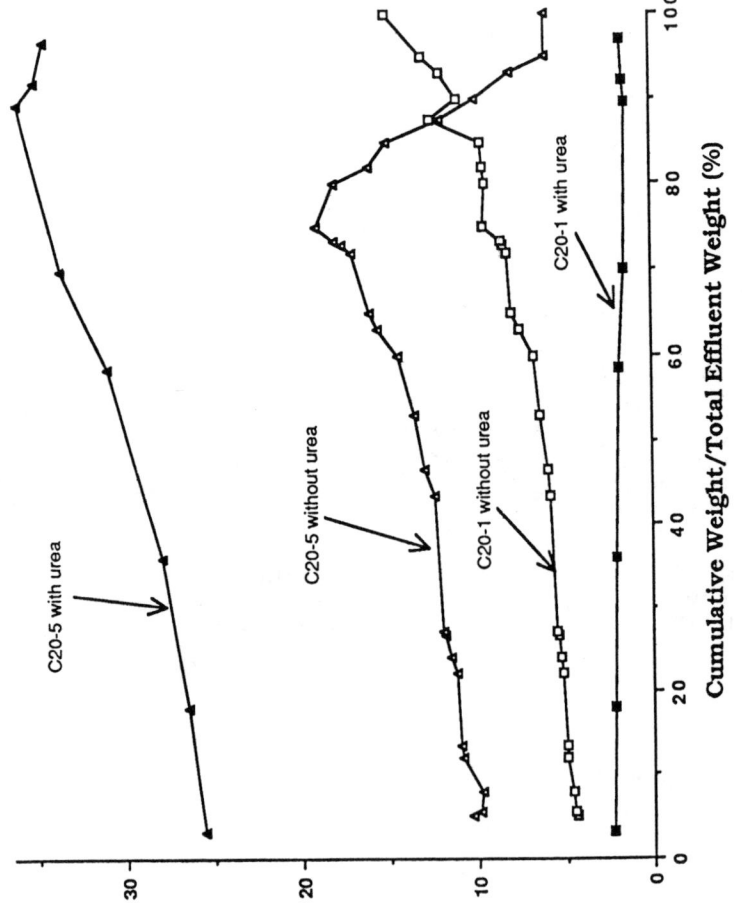

Figure 10. Supercritical CO_2 extraction of C20 methyl esters with and without urea (Reproduced with permission from Ref. 18. Copyright 1986 Gakkai Shuppan Senta.)

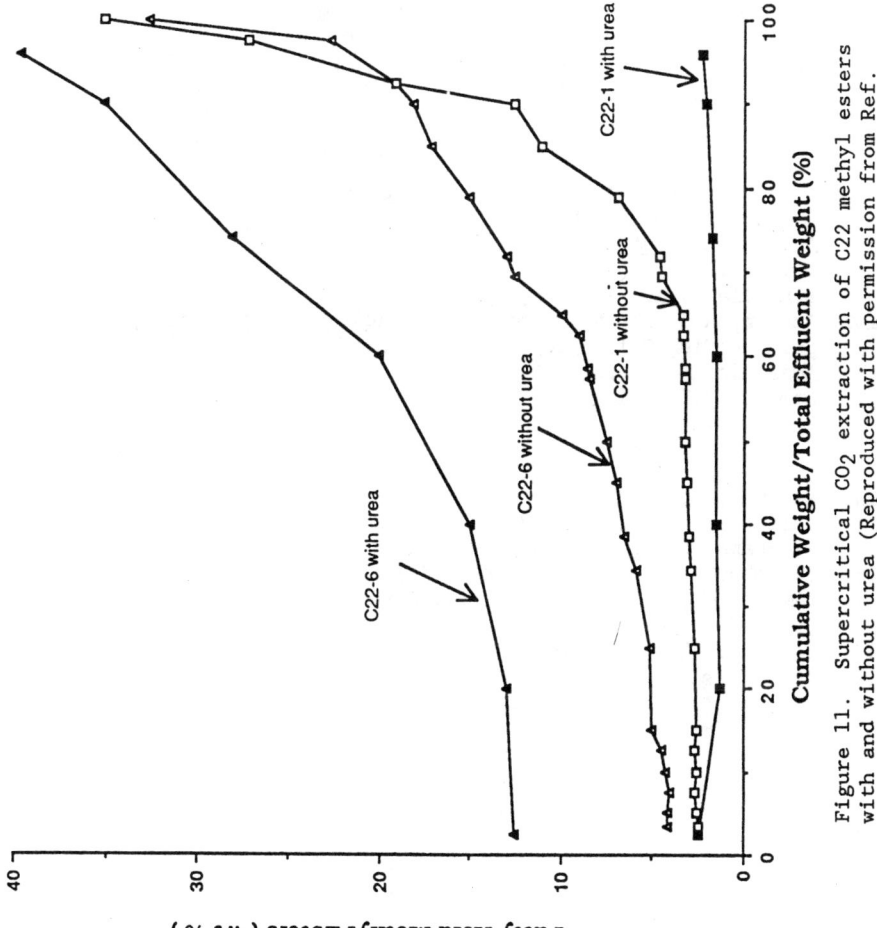

Figure 11. Supercritical CO_2 extraction of C22 methyl esters with and without urea (Reproduced with permission from Ref. 18. Copyright 1986 Gakkai Shuppan Senta.)

Literature Cited

1. *Yearbook of Fishery Statistics*, Food and Agriculture organization of the United Nations: Rome, 1984, 61.
2. *Processed Fisheries Products, Annual Summary*, Current Fisheries Service, NOAA, U.S. Department of Commerce: Washington, DC, 1985.
3. Dyerberg, H.M.; Bang, H.O. *Lancet* ii 1979, p 433.
4. Sinclair, H.M. *Postgrad. Med. J.* 1980, 56, 579.
5. Kremer, J.M.; Bigauotette, J.; Michalek, A.V.; Timchalk, M.A.; Linnger, L.; Rynes, R.I.; Huyck, C.; Zieminski, J.; Bartholomew, L.E. *The Lancet* 1985, 1 (8422), 184.
6. Virage, R.; Bouilly, P.; Fryman, D. *The Lancet* 1985, 1 (8422), 181.
7. Kinsella, J.E. *Seafoods and Fish Oils in Human Health and Disease* Marcel Dekker: N.Y., 1987.
8. Anonymous. *Tufts University Diet & Nutrition Letter* 1987, 4 (11), 2.
9. Kinsella, J.E. *Food Technology* 1986, 2, 89.
10. Ackman, R.G. In *Nutritional Evaluation of long-chain Fatty acid in Fish Oil*; Barolow, S.M.; Stansby, M.E., Ed.; Academic: New York, 1982; p 25.
11. Bonnet, J.E.; Sidwell, V.D.; Zook, E.G. *Fish. Rev.* 1974, 36(2), 8.
12. Gruger, E.G.; Nelson, Jr., R.W.; Stansby, M.E. *J. Am. Oil. Chem. Soc.* 1964, 41(10), 117.
13. Randall, L.G. *Sci. Tech.* 1982, 17, 1.
14. Saito, S. *Petrotech.* 1982, 5, 115.
15. Nagahama, K. *Bunrigizyutsu.* 1981, 11, 23.
16. Eisenbach, W. *Ber. Bunsenges. Phy. Chem.* 1984, 88, 882.
17. Domart, C.; Miyauchi, D.T.; Sumerwell, W.N. *J. Am. Oil. Chem. Soc.* 1955, 32(9), 481.
18. Saito, S. *Kagaku to Seibutsu* 1986, 24, 201.
19. Krukonis, V.J. *Presented at the 75th Annual Meeting of the American Oil Chemists' Society*, Dallas, April 29-May 3, 1984.
20. Nilsson, W., et al. *Presented at the 77th Annual Meeting of the American Oil Chemists' Society*, Honolulu, May 19, 1986.
21. Daniels, J.A., et al. "Concentration of n-3 Fatty Acids from Fish Oil Using Supercritical CO_2"; Internal Report; Cornell University: Ithaca, NY, 1986.

RECEIVED October 28, 1987

Chapter 6

Supercritical Carbon Dioxide Extraction of Terpenes from Orange Essential Oil

F. Temelli [1], R. J. Braddock [1], C. S. Chen [1], and S. Nagy [2]

[1] Citrus Research and Education Center, Institute of Food and Agricultural Sciences, University of Florida, Lake Alfred, FL 33850
[2] Department of Citrus, Citrus Research and Education Center, University of Florida, Lake Alfred, FL 33850

> This report reviews the recovery, composition and folding of cold-pressed orange oil and discusses the application of supercritical fluid extraction technology in this field. Supercritical carbon dioxide extraction is a suitable low-temperature alternative process to distillation for separating terpene hydrocarbons from the oxygenates in cold-pressed citrus oils. Citrus essential oils are important flavoring agents for the food industry. The oxygenated compounds are responsible for the characteristic citrus flavor. The terpene hydrocarbons are the major components; however, in some applications they are detrimental to flavor and aroma. Terpene hydrocarbons are unsaturated compounds and are readily decomposed by heat, light and oxygen. To obtain a more stable product, some terpenes are generally removed by distillation. However, heating during distillation results in development of some off-flavors.

Matthews and Braddock (1) have estimated a potential yield in Florida of 13,800 tons of orange oil for the 1984-85 season. However, it has been estimated that actual cold-pressed orange oil production is only about 20% of the potential with another 25-30% recovered as d-limonene (2).

Citrus oils are used in a wide variety of applications in many major industries which include flavor, beverage, food, cosmetics, soap, pharmaceutical, chemical and insecticide. The most important outlet is the flavor industry. The major component, d-limonene, is used for the manufacture of materials like spearmint oil flavor, l-carvone, terpene resins and adhesives.

Citrus oil is present in small glands contained in the flavedo, which is the colored portion of the peel of the fruit. Cold-pressed citrus peel oil is obtained commercially by a process that starts with the rupture of these glands during juice extraction. For its recovery, the oil is washed away from the peel with water forming an

0097-6156/88/0366-0109$06.00/0
© 1988 American Chemical Society

oil-water emulsion. It is important to maintain an excess of water to prevent the oil from being reabsorbed by the peel once it is released. A two-stage centrifugation process is used to recover orange oil from the oil-water emulsion. The first stage is called a desludger and its output contains 60-80% oil. The polisher is the second stage centrifuge and continuously discharges the pure orange oil and intermittently discharges an aqueous waste. Cold-pressed oil is winterized at -29°C, 3-5 days or at -4°C, 4-5 weeks to remove waxes which could produce a cloudy appearance in the final product. Dewaxed oils should be stored at 15-22°C, under an inert gas atmosphere. Kesterson and Braddock (3), and Kesterson et al. (4) have discussed the production, physical properties of the citrus oils, and the parameters affecting them.

Composition of Peel Oils

Citrus, like many other essential oils, consists of mixtures of hydrocarbons of the 'terpene' and 'sesquiterpene' groups, oxygenated compounds and nonvolatile residues. Terpenes make up approximately 95% by weight of orange oils while the oxygenated compounds constitute the remaining 5%. The compositions of citrus oils have been studied extensively by different researchers (5-9). Citrus terpenes ($C_{10}H_{16}$) are mostly unsaturated acyclic and cyclic compounds derived from the condensation of two 5-carbon isoprene units. Sesquiterpenes have the general formula $C_{15}H_{24}$. Advances in instrumental analysis, e.g. gas chromatography and mass spectroscopy, have made it possible to identify more than 150 compounds in cold-pressed citrus oils. Shaw (10-11) has extensively reviewed the qualitative and quantitative analyses of citrus essential oils.

The terpene fraction, made up mainly of limonene has little contribution to the flavor or fragrance of the oil. Moreover, since its make up is mostly unsaturated compounds, this fraction exhibits instability to heat and light, and is rapidly oxidized by atmospheric oxygen, consequently decomposing to undesirable compounds. The mechanism of oxidation of limonene and the factors affecting it have been studied by Buckholz and Daun (12), Bernhard and Marr (13) and Newhall and Kesterson (14).

Oxygenated compounds are highly odoriferous and are the principal odor impact compounds. This flavor fraction consists of aldehydes, alcohols, ketones, esters, ethers and phenols. The total aldehyde content of oils is used as an indicator of quality and this is measured as decanal since it is often the major aldehyde. Aldehyde concentration in orange oil is generally about 1.5%. The two major aldehydes in orange oil, octanal and decanal, have been quantitated as 0.2-2.8% and 0.1-0.7%, respectively, by different workers (10).

The alcohols in orange oil which are most important to flavor are linalool, α-terpineol and terpinen-4-ol. Reported values for linalool range from 0.3-5.3% of cold-pressed oil. Both α-terpineol and terpinen-4-ol are degradation products of d-limonene and could be formed if the peel oil is allowed to remain in contact with the acidic juice for any length of time during processing. A microbial degradation product of limonene, α-terpineol, contributes to off-flavor in stored orange juice (15).

Ketones quantitated in orange oil are carvone, an oxidation product of limonene, found at 0.1% or less, and nootkatone, found at extremely low levels of < 0.01% of peel oil.

Folding Processes

The industrial practice of folding is to remove some of the limonene along with other unstable terpenes and to concentrate the oxygenated compounds. This process of folding the oil results in a more stable product. Five- and ten-fold oils which are concentrated to one-fifth or one-tenth of original weight are the most common products. Higher concentrations lose the natural character of the citrus due to absence of terpenes which do, nonetheless, contribute some flavor notes. An advantage of folding is the reduction of water insoluble terpenes which may produce an unaesthetic 'ring' at the neck of the soft drink bottles evident when unfolded cold-pressed oil is used. Folded oils are also preferred in flavoring food products that will be heated and in beverages where limonene and oxidation products might be objectionable. An obvious advantage of folded oils is a reduction in storage and transportation costs due to reduced volumes.

Vacuum distillation, steam distillation, extraction with solvents, and adsorption processes are used for folding cold-pressed oils. How each affects the quality of the folded oil will be briefly described.

In Table I, the boiling range, at atmospheric pressure, of terpenes varies from 150 to 180°C and that of sesquiterpenes between 240 and 280°C; the boiling points of most oxygenated compounds lie between those of terpenes and sesquiterpenes. This difference in boiling points is used to separate them by distillation. However, the temperature must be kept as low as possible with the aid of vacuum (16). Time, temperature and vacuum have major effects on the quality and yield. Losses occur in the aldehyde flavor fraction as the level of folding increases (17). Specific gravity, refractive index, optical rotation, aldehyde value, ester value and gas-liquid chromatogram response change linearly with concentration during vacuum distillation of citrus oils and this could be used as an objective evaluation of the degree of folding (18). Vora et al. (19) concentrated orange oil by vacuum distillation (57-62°C; 1.3 kPa), and gave a quantitative analysis of the constituents. During steam distillation orange oil is exposed to a relatively high temperature which can lead to artifacts due to the hydrolytic influence of water.

In the process of extraction with low boiling organic solvents, an important consideration is the presence of residual solvent in the extract. Also, extraction with organic solvents does not totally eliminate thermal degradation since the solvent has to be removed by distillation. Owusu-Yaw et al. (20) extracted cold-pressed Valencia orange oil with aqueous ethyl alcohol to remove the terpenes and sesquiterpenes. The ratio 1:3 of oil:solvent resulted in sesquiterpeneless oils with a low terpene content. However, this ratio also gave low oil recoveries.

Table I: Formula, Molecular Weight, and Boiling Point of Compounds Present in Cold-Pressed Orange Oil

Compound	Peak No.	Formula	Molecular Weight	Boiling Point(°C)
α-pinene	1	$C_{10}H_{16}$	136	155
Sabinene	2	$C_{10}H_{16}$	136	149
Octanal	3	$C_8H_{16}O$	128	172
Myrcene	4	$C_{10}H_{16}$	136	166
Phellandrene	5	$C_{10}H_{16}$	136	171
d-limonene	6	$C_{10}H_{16}$	136	177
1-octanol	7	$C_8H_{18}O$	130	195
Nonanal	8	$C_9H_{18}O$	142	185
Linalool	9	$C_{10}H_{18}O$	154	209
Citronellal	10	$C_{10}H_{18}O$	154	203
α-terpineol	11	$C_{10}H_{18}O$	154	220
Decanal	12	$C_{10}H_{20}O$	156	208
Neral	13	$C_{10}H_{16}O$	152	224
Geranial	14	$C_{10}H_{16}O$	152	230
Perillaldehyde	15	$C_{10}H_{14}O$	150	236
Undecanal	16	$C_{11}H_{22}O$	170	117
Dodecanal	17	$C_{12}H_{24}O$	184	231
β-caryophyllene	18	$C_{15}H_{24}$	204	288
β-copaene	19	$C_{15}H_{24}$	204	119
β-farnesene	20	$C_{15}H_{24}$	204	121
Valencene	21	$C_{15}H_{24}$	204	
β-sinensal	22	$C_{15}H_{22}O$	218	
α-sinensal	23	$C_{15}H_{22}O$	218	
Nootkatone	24	$C_{15}H_{22}O$	218	

Terpeneless and sesquiterpeneless oils have been prepared by adsorption on silicic acid which is followed by eluting the terpenes with hexane and oxygenated compounds with ethanol (21-23).

The drawbacks of all the processes discussed are low yields, formation of degradation products, and/or the removal of solvents.

Supercritical Fluid Extraction

Supercritical fluid extraction processes have gained increasing importance in the chemical and food industries in recent years, since in many applications product recovery and quality can be maximized while minimizing energy requirements. Even though scientists, since the late 1800's, knew the enhanced solvating power of many gases above their critical points, it was not until recently that supercritical fluids have been the focus of active research. Consequently, a large number of review papers have been published (24-29), and the feasibility of different applications of the process in the food industry has been investigated. Some of these applications are decaffeination of coffee and tea (27-29), extraction of oil from vegetable seeds (30-34), removal of nicotine from tobacco (35), and production of hop (35-37) and spice (35) extracts. Increased cost of energy, tightening government regulations on solvent residues and increased demand for higher quality products are some of the motivations given for utilizing supercritical fluid technology. The availability of inexpensive energy, poor understanding of the thermodynamics required to describe or predict such processes, high capital costs associated with plant start-up and operation, and an absence of engineering data for scale-up technology are factors that have been cited to explain the limited acceptance of supercritical processes to this date.

Supercritical extraction provides some distinct advantages over other separation techniques: thermally unstable compounds can be separated at low temperatures; the solvent can be removed easily from the solute by reducing the pressure and/or adjusting the temperature; thermal energy requirements are lower than that for distillation; surprisingly high selectivities for the solute can be accomplished; rapid extraction can be achieved due to low viscosity, high diffusivity and good solvating power of the supercritical fluid solvent. Carbon dioxide as a supercritical fluid solvent is attractive since it is nontoxic, nonflammable, inert, readily available in high purity, inexpensive, has low surface tension and viscosity, and high diffusivity.

Factors Affecting Solubility in Supercritical Fluids

Solute Physical Properties. Vapor pressure, polarity and molecular weight are the most important factors affecting the solubility of compounds in a supercritical fluid. Vapor pressure versus temperature curves of some cold-pressed orange oil components are given in Figure 1 (data from (38)). The terpene hydrocarbons have the highest vapor pressures. Vapor pressure of limonene is 4.3 times greater than that of linalool at 50°C, and 4.7 times at 40°C. Linalool is the compound with the highest vapor pressure among the

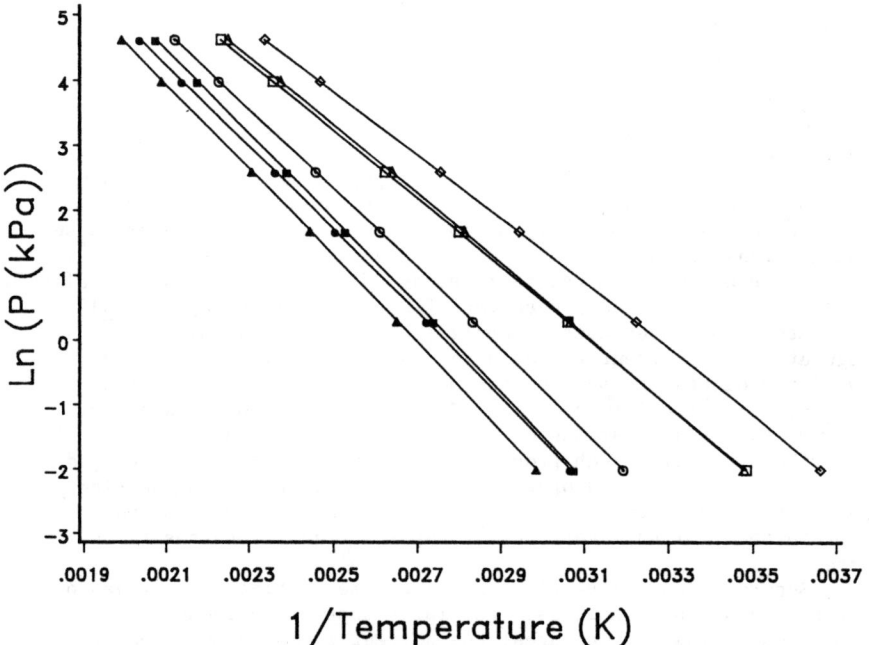

Figure 1. Vapor pressure vs. temperature curves of important components of orange essential oil: (◇-◇) α-pinene, (□-□) d-limonene, (Δ - Δ) myrcene, (o - o) linalool, (● - ●) α-terpineol, (■-■) decanal, (▲ - ▲) α-citral (Data from (38)).

oxygenated compounds. Other properties of these compounds are
summarized in Table I. The most abundant terpene hydrocarbons are
α-pinene, sabinene, myrcene and d-limonene. They all have a
molecular weight of 136. Sesquiterpene hydrocarbons include
β-caryophyllene, β-copaene, β-farnesene and valencene. Their
molecular weight is 204. Oxygenated flavor compounds include
aldehydes (octanal, nonanal, citronellal, decanal, neral, geranial,
perillaldehyde, undecanal, dodecanal, α- and β-sinensal), alcohols
(1-octanol, linalool, α-terpineol) and ketones (nootkatone). Their
molecular weights are between 152 and 156. Since carbon dioxide has
no dipole moment and has a greater affinity for nonpolar solutes,
terpenes which are nonpolar, have less molecular weight and a higher
vapor pressure, would be more soluble in supercritical carbon
dioxide than the oxygenated flavor fraction.

Temperature and Pressure. Solvent power of a supercritical fluid is
directly related to its density. In the vicinity of the critical
point, large density changes can be produced with either relatively
small pressure or temperature changes. At a given temperature, the
solvent power or the fluid density increases as the pressure is
increased. A rise in temperature at constant pressure leads to a
decrease in solvent density. This strong pressure dependence of the
dissolving power of a supercritical fluid is exhibited by all solid
and many liquid solutes as long as the solute is not infinitely
miscible with the solvent. This pressure dependency is a key factor
in supercritical fluid extraction. Stahl and Gerard (39) have
studied the solubility behavior of certain essential oil components
in supercritical carbon dioxide. As noted from Figure 2, all
components exhibited some solubility in carbon dioxide at low
densities. With increasing pressure above the critical pressure of
carbon dioxide (7.4 MPa), the gas phase concentration of the solutes
increased exponentially. The temperature of the system was 40°C
which is just above the critical temperature of pure carbon dioxide
(31°C).

The effect of temperature on the solubility of a substance in a
supercritical fluid changes with pressure. At pressures close to
the critical point a temperature rise results in a decrease in the
concentration of the solute in the supercritical phase. However, at
high pressures, a rise in temperature causes an increase in the
solubility because a rise in temperature at constant pressure leads
to a decrease in gas density, and at the same time results in an
increase in the vapor pressure of the solute. The reduction in gas
density, due to an increase in temperature, becomes less at higher
pressures than at low pressures. So, the increase in vapor pressure
of the solute overcomes the decrease in gas density and leads to a
higher concentration in the supercritical phase.

Potential Citrus Oil Applications

Applications of supercritical fluid extraction to essential oil
processing have been described in recent literature references
(39-43). However, the literature lacks detailed data on
multicomponent citrus essential oils with supercritical carbon
dioxide. Stahl and Gerard (39) used pure essential oil components

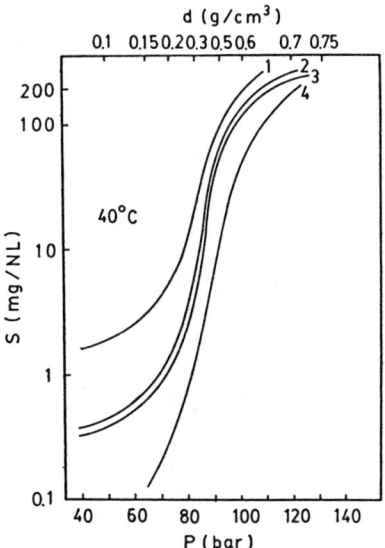

Figure 2. Solubility isotherms of essential oil components in dense carbon dioxide at 40°C; (1) limonene, (2) carvone, (3) caryophyllene, (4) valeranone (Reproduced with permission from Ref. 39. Copyright 1985 S. Allured).

to study their solubility in dense carbon dioxide. They showed that the separation of sesquiterpene hydrocarbons and oxygenated monoterpenes is difficult since their solubility behavior and vapor pressures are almost the same and made this separation possible by using their differences in polarity and saturating the carbon dioxide with water. Stahl and Gerard (39) concluded that fractionation is possible at 7-8 MPa and 40°C in that monoterpenes, monoterpene derivatives and sesquiterpene hydrocarbons can be extracted with dry supercritical CO_2 leaving oxygenated sesquiterpenes behind. Coppella and Barton (40) studied the vapor-liquid equilibrium for lemon oil and carbon dioxide at 30-40°C and 4-9 MPa. They used the Peng-Robinson equation of state, treating the system as CO_2:limonene binary with temperature-dependent interaction parameter to model the system. They determined the relative volatilities of terpinolene to citronellal as 2 and limonene to geranial as 5. Robey and Sunder (41) have used the supercritical carbon dioxide extraction technique to separate lemon oil into its fractions. The terpene fraction was preferentially solubilized at all operating conditions tested. They studied the effects of varying the extraction temperature between 40 and 90°C and determined that yield increased. The same trend was true for the separation factor of limonene/citral. Fitting an equation of state to the data showed that those temperature-pressure regions giving high total solubility also tend to give poor selectivity.

In general, terpene hydrocarbons can be extracted from cold-pressed citrus oils under conditions close to the critical point of carbon dioxide. Temperatures should be within the range 31 to 70°C, since citrus oils begin to suffer thermal degradation above this limit. Extraction pressures should be between 7.4 to 13.0 MPa; since at high pressures the increased density of carbon dioxide increases solubility of the oxygenated compounds making fractionation impossible. Another factor in determining the extraction pressure is related to economic considerations, since construction and operating costs increase with higher system pressures. Gerard (42) described a continuous, countercurrent high-pressure extraction column with built-in baffles to enlarge the surface area and a reflux of the top product. He proposed an extraction pressure of 8 MPa and temperature of 60 to 70°C which would give the advantage of reaching a higher loading of the gas phase and the possibility of creating a reflux by cooling the column head. Robey and Sunder (41) used a computer program to simulate a multistage column to produce a ten-fold concentration of oxygenated compounds in lemon oil and showed that a countercurrent column operating at 10.3 MPa and 60°C, with 12 stages and reflux would produce the desired product at 99% yield.

Extraction System

The Supercritical Extraction Screening System with a 300 mL extraction chamber designed by the Autoclave Engineers, Inc. (Erie, Pennsylvania) was used (Figure 3). After some modifications, the system could be used both as a static cell and as a dynamic flow

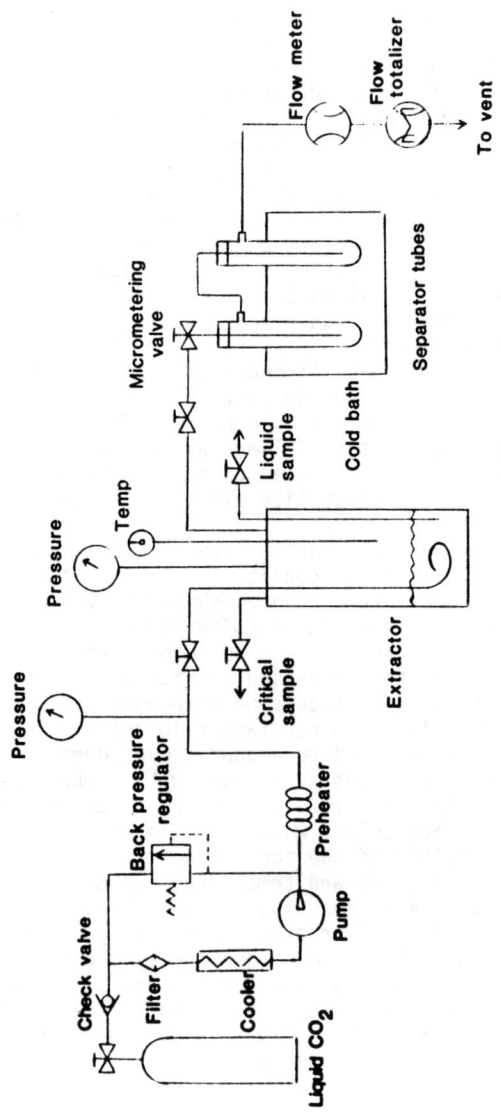

Figure 3. Schematic of the supercritical carbon dioxide extraction system.

apparatus. The details of each system, advantages and disadvantages were outlined by McHugh and Krukonis (44).

Static Mode. The system can be used as a static cell to obtain equilibrium data for supercritical carbon dioxide and cold-pressed Valencia orange oil. Using the system as a static cell to obtain phase equilibrium data under supercritical conditions was complicated by two factors: 1) trapping small samples without disturbing the equilibrium is difficult and 2) the small sample size makes subsequent analysis complex. The procedure was as follows. The 20 mL oil sample was placed into the extractor and carbon dioxide was pumped until the desired pressure was reached. At this time, the extractor was isolated by closing the valves at the inlet and outlet lines. Extraction time of 5 hours was enough for the orange oil-supercritical carbon dioxide system to reach equilibrium with agitation under constant temperature and pressure. When the liquid and supercritical phase samples were taken from the extractor, the temperature of the system did not change and the drop in pressure was about 5%. Liquid phase samples were taken from the sampling valve on the extractor whose inlet tube extends to the bottom (Figure 3). Vapor phase samples were filled into an 8-mL evacuated sampling bomb from the vapor sampling valve on the extractor. Components of the oil that were solubilized by carbon dioxide were condensed in a separator which was in a liquid nitrogen bath while depressurizing the supercritical phase sample by slowly opening the valve on the sampling bomb (Figure 4). After depressurizing, the sampling bomb was heated and a slight vacuum was applied to make sure that all the sample was removed from the bomb and the sample was collected in the separator tube.

Dynamic Mode. Dynamic or flow techniques are used for determining solute solubilities in a supercritical fluid and also for stripping and fractionation studies. This technique allows one to obtain equilibrium, stripping, or fractionation data rapidly and reproducibly. The system operated as follows under the dynamic mode. Cold-pressed Valencia orange oil samples were placed in a 300 mL extraction chamber and the temperature adjusted with the use of heaters around the chamber. The high-pressure pump delivered carbon dioxide continuously at flow rates slow enough to insure that equilibrium was obtained between the oil and the carbon dioxide as it flowed through the system. Pressure was adjusted by a back pressure regulator. Typical equilibrium flow rates were reported to be 60-500 mL/min as measured at 100 kPa and 0°C (44). Carbon dioxide flowed through a preheater to insure that it reached extraction temperature before contacting the oil.

As carbon dioxide entered the extractor, it distributed through the oil. Some of the carbon dioxide was solubilized in the oil resulting in the expansion of the liquid phase. At the same time carbon dioxide solubilized some of the oil components and carried them out of the extractor. Pressure was reduced through an expansion valve which was heated to prevent freezing. Extracts were collected in a glass tube (trap) placed in a -25°C propylene glycol bath. The amount of oil collected in the cold trap for a given amount of time was determined gravimetrically and the corresponding

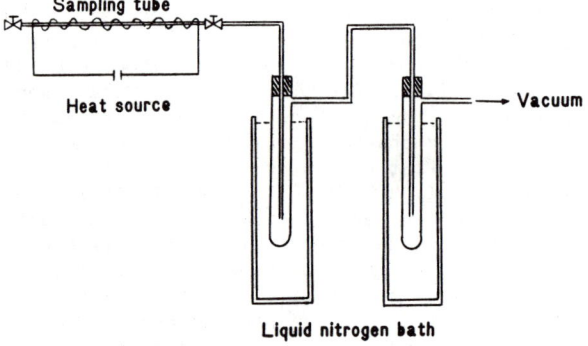

Figure 4. Static mode critical phase sampling system.

volume of CO_2 was measured with a wet gas meter. The instantaneous flow rate of the expanded CO_2 was measured with a bubble-flow meter. The extract collected and oil left behind were analyzed by gas-liquid chromatography and the components identified by mass spectrometry and comparison with standards.

Results and Discussion

Gas-liquid chromatograms of the liquid and supercritical phase samples at equilibrium under 8.3 MPa pressure and at 70°C are given in Figure 5, and identification of the peaks are specified in Table I. The first six peaks are terpene hydrocarbons except for octanal which elutes with the terpenes in the GC analysis. In order to obtain a separation of the terpenes from the flavor compounds, it is necessary to have a higher concentration of the terpenes in the supercritical phase. As can be seen from the chromatograms, the terpenes were solubilized by supercritical carbon dioxide to a greater extent, and the oxygenated compounds were left in the liquid phase. These chromatograms show that the peak areas for the terpenes were more than doubled and decreased approximately by half for the flavor compounds in the supercritical phase when compared to the liquid phase. The ratio of the oxygenated compounds to terpenes was 0.0084 and 0.0224 in the critical and liquid phases, respectively.

The results of continuous flow experiments are presented in Figure 6 as the amount of extract as a function of time. Cold-pressed Valencia orange oil (20 mL) was extracted with supercritical carbon dioxide for five hours and the oil extract was collected and analyzed after every hour. All four runs were performed at 8.3 MPa. The amount of extract collected followed a linear trend over time. After five hours at 70°C, the amount of extract was 0.26 at 50 mL/min and 1.31 g at 500 mL/min of carbon dioxide (flow measured at 100 kPa and 25°C). When the flow rate was increased 10 times, the amount of extract was increased by about 5 times. Even though a smaller amount of oil was collected at a low flow rate, the concentration of oil in carbon dioxide was twice that at higher flow rate. This was due to increased contact time between carbon dioxide and oil at low flow rates.

These extractions were also carried out at 40°C. When the temperature was decreased from 70 to 40°C at 8.3 MPa, two changes occured: first, the density of carbon dioxide increased from 0.18 to 0.32 g/cm^3, which resulted in an increase in the dissolving power of carbon dioxide, and secondly, the vapor pressure of limonene decreased from 2.86 to 0.64 kPa. Since the amounts of extract collected under each condition were very close, a conclusion cannot be drawn as to which factor has a greater effect. However, the effect of temperature and pressure on solubility was studied in detail and will be discussed in another paper. The gas chromatographic analyses of the extract and the oil residue resembled those given in Figure 5. It should be noted that the qualitative analysis of the feed oil was similar to the oil residue with a slight increase in the compositions of the oxygenated compounds because 1.5% and 7.8% of the feed oil was removed as extract after 5 hours at 50 and 500 mL/min flow rate, respectively.

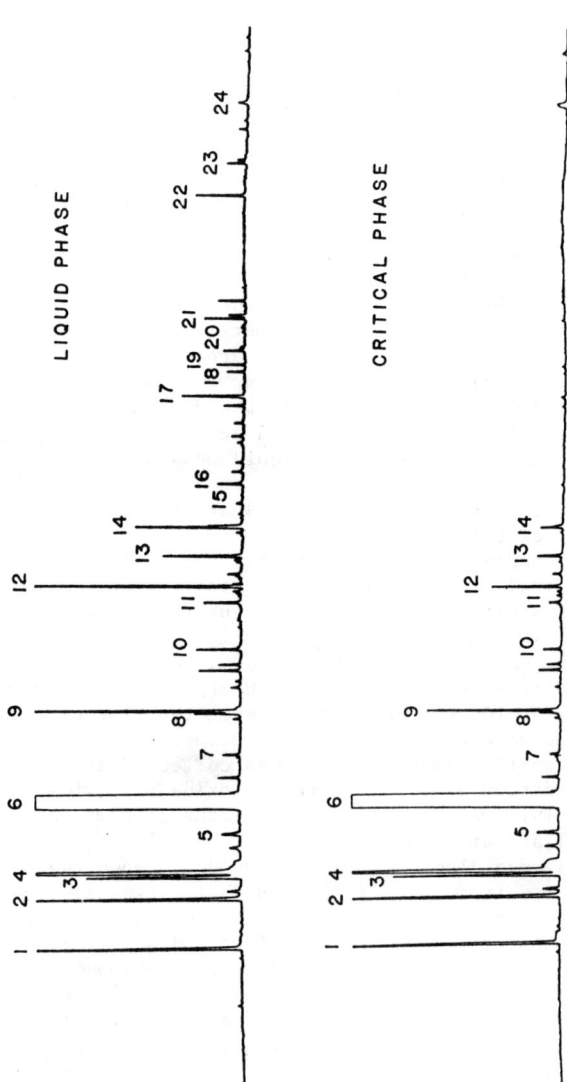

Figure 5. Gas-liquid chromatograms of orange oil liquid and supercritical phase samples at equilibrium at 70 °C and 8.3 MPa. GC conditions: 60 m SE-30 capillary column with FID detector. Injector at 200 °C, detector at 250 °C, column temperature programmed from 75 to 200 °C with a ramp rate of 6 degrees/min., with initial and final holding times of 11 and 10 min., respectively. Split ratio for the injection port 100:1. 0.2 μl sample injections. Carrier gas is hydrogen at a pressure of 138 kPa.

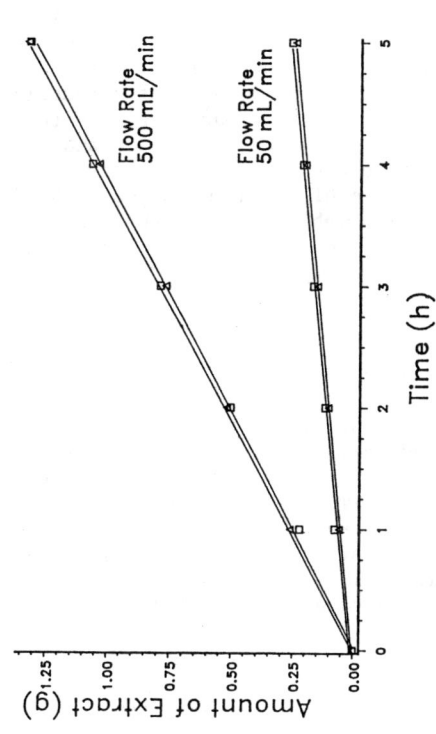

Figure 6. Amount of oil extracted as a function of time at 8.3 MPa and temperatures of 70°C (□-□) and 40°C (Δ-Δ) at flow rates of 50 and 500 mL/min.

Although the amount of extract was linear in the 5 hour period, it probably will not be over a longer period of time. In a batch system, since the composition of the feed oil will change with time, the amount of oxygenated compounds in the extract also will change. Based on these assumptions, it was found that at least 50 hours of extraction would be necessary to obtain a 5-fold oil at a flow rate of 500 mL/min. However, leaving the oil at 70°C for 50 hours would probably result in the formation of undesirable terpene degradation products. Therefore, no attempt was made to produce a 5-fold oil with the system described.

In conclusion, it is possible to concentrate the flavor fraction of cold-pressed citrus oils with supercritical fluid technology by selectively extracting the terpenes from the oil. During continuous extractions, the amount of extract followed a linear trend with time over the first 5 hours of extraction and it increased five times when the flow rate was increased ten times. Since the design of supercritical fluid extraction and solvent regeneration processes for the concentration of citrus oils require accurate calculation of phase equilibria, more research must be done to determine the equilibrium solubility data, the thermodynamic model to represent the system, and the economic feasibility of the process.

Literature Cited

1. Matthews, R. F.; Braddock, R. J. Food Technol. 1987, 41(1), 57.
2. Braddock, R. J.; Miller, W. M. J. Food Sci. 1982, 47, 2008.
3. Kesterson, J. W.; Braddock, R. J. "By-Products and Specialty Products of Florida Citrus", Univ. of Fla. Agric. Exp. Stn. Tech. Bull. 784: Gainesville, 1976, 122 pp.
4. Kesterson, J. W.; Hendrickson, R.; Braddock, R. J. "Florida Citrus Oils", Univ. of Fla. Agric. Exp. Stn. Tech. Bull. 749: Gainesville, 1971, 180 pp.
5. Braddock, R. J.; Kesterson, J. W. J. Food Sci. 1976, 41, 1007.
6. Shaw, P. E.; Coleman, R. L. J. Agr. Food Chem. 1974, 22(5), 785.
7. Berry, R. E.; Shaw, P. E.; Tatum, J. H.; Wilson III, C. W. Food Technol. 1983, 37(12), 88.
8. Braddock, R. J.; Kesterson, J. W. Proc. Fla. State Hort. Soc. 1976, 89, 196.
9. Shaw, P. E.; Moshonas, M. G. Mass Spectrom. Rev., 1985, 4, 397.
10. Shaw, P. E. J. Agr. Food Chem. 1979, 27, 246.
11. Shaw, P. E. In "Citrus Science Technology"; Nagy, S.; Shaw, P. E.; Veldhuis, M. K., Eds.; AVI: Westport, 1977; Vol. 1, pp. 427-462.
12. Buckholz, L. L.; Daun, H. K. J. Food Sci. 1978, 43, 535.
13. Bernhard, R. A.; Marr, A. G. Food Res. 1960, 25, 517.
14. Newhall, W. F.; Kesterson, J. W. Proc. Fla. State Hort. Soc. 1961, 73, 239.
15. Tatum, J. H.; Nagy, S.; Berry, R. E. J. Food Sci. 1975, 40, 707.

16. Guenther, E. "The Essential Oils", 2nd ed.; Robert E. Krieger Publishing Co.: New York, 1972; Vol. 1, pp. 218.
17. Braddock, R. J. In "Citrus Nutrition and Quality"; Nagy, S.; Attaway, J. A., Eds.; ACS Symposium Series, No. 143, 1980, pp. 273-288.
18. Lifshitz, A.; Sterak, Y.; Elroy, I. Perf. & Ess. Oil Rec. 1969, 60, 157.
19. Vora, J. D.; Matthews, R. F.; Crandall, P. G.; Cook, R. J. Food Sci. 1983, 48, 1197.
20. Owusu-Yaw, J.; Matthews, R. F.; West, P. F. J. Food Sci. 1986, 51, 1180.
21. Kirchner, J. G.; Miller, J. M. Ind. Eng. Chem. 1952, 44(2), 318.
22. Lijn, J.; Lifshitz, A. Lebensm.-Wiss. u. Technol. 1969, 2, 39.
23. Ferrer, O. MS thesis, University of Florida, Gainesville, FL., 1984.
24. Rizvi, S. S. H.; Daniels, J. A.; Benado, A. L.; Zollweg, J. A. Food Technol. 1986, 40(7), 57.
25. Rizvi, S. S. H.; Benado, A. L.; Zollweg, J. A.; Daniels, J. A. Food Technol. 1986, 40(6), 55.
26. Williams, D. F. Chem. Eng. Sci. 1981, 36, 1769.
27. Paulaitis, M. E.; Krukonis, V. J.; Kurnik, R. T.; Reid, R. C. Rev. Chem. Eng. 1983, 1(2), 179.
28. Zosel, K. In "Extraction with Supercritical Gases"; Schneider, G. M.; Stahl, E.; Wilke, G., Eds.; Verlag Chemie: Deerfield Beach, 1980; pp. 1-23.
29. Caragay, A. B. Perfumer & Flavorist Aug/Sep 1981, 6, 43.
30. Eldridge, A. C.; Friedrich, J. P.; Warner, K.; Kwolek, W. F. J. Food Sci. 1986, 51, 584.
31. Friedrich, J. P.; Pryde, E. H. J. Am. Oil Chem. Soc. 1984, 61(2), 223.
32. Bulley, N. R.; Fattori, M.; Meisen, A.; Moyls, L. J. Am. Oil Chem. Soc. 1984, 61(8), 1362.
33. List, G. R.; Friedrich, J. P. J. Am. Oil Chem. Soc. 1985, 62(1), 82.
34. Stahl, E.; Schutz, E.; Mangold, H. K. J. Agric. Food Chem. 1980, 28(6), 1153.
35. Hubert, P.; Vizthum, O. G. Angew. Chem. Int. Ed. Engl. 1978, 17, 710.
36. Harold, F. V.; Clarke, B. J. Brewer's Digest Sep. 1979, pp. 45.
37. Grimmet, C. Chem. Ind. May 1981, 359.
38. Weast, R. C.; Astle, M. J.; Beyer, W. H., Eds. "Handbook of Chemistry and Physics", 67th ed.; CRC Press, Inc.: Boca Raton, 1986.
39. Stahl, E.; Gerard, D. Perfumer & Flavorist Apr/May 1985, 10, 29.
40. Coppella, S. J.; Barton, P. In "Supercritical Fluids: Chemical Engineering Principles and Applications"; Squires, T. G.; Paulaitis, M. E., Eds.; ACS Symposium Series, No. 329, 1987, pp 202-212.

41. Robey, R. J.; Sunder, S., presented in part at the annual meeting of the American Institute of Chemical Engineers, San Fransisco, 1984.
42. Gerard, D. Chem. Ing. Tech. 1984, 56, 794.
43. Stahl, E.; Quirin, K. W.; Glatz, A.; Gerard, D.; Rau, G. Ber. Bunsenges. Phys. Chem. 1984, 88, 900.
44. McHugh, M.; Krukonis, V. J. "Supercritical Fluid Extraction - Principles and Practice", Butterworth Publishers: Boston, 1986.

RECEIVED October 9, 1987

Chapter 7

Steps To Developing a Commercial Supercritical Carbon Dioxide Processing Plant

R. T. Marentis

Supercritical Fluid Processing Systems, Pitt-Des Moines, Inc., 35 Airport Road, Morristown, NJ 07960 and Neville Island, Pittsburgh, PA 15225

> Commercialization of a supercritical carbon dioxide processed food product requires the successful application of five sequential steps: 1) application of high pressure CO_2 phase equilibria and fluid dynamics theory; 2) knowledge of the botanicals structure and chemistry; 3) performance of the "process design protocol"; 4) preliminary process design and economic evaluation; and 5) design, construction and start-up of the commercial-scale plant. Many decisions made during the early steps have large impacts on commercial plant performance capabilities and economic efficiency. The impact on commercial plant performance and economics should be factored into decisions made at every commercialization step.

Supercritical fluid ($SCFCO_2$) processing is of increasing commercial importance to the food industry in Europe and America. Citations of $SCFCO_2$ extractions of botanicals are replete in the literature. F. M. Taylor (1) lists over 100 literature citations on herb and spice applications alone. In general, $SCFCO_2$ extracts have no solvent residues, no off-odors from "still notes", higher concentrations of the most valuable components (due to the extraordinary selectivity of the $SCFCO_2$ solvent) when compared with conventional organic liquid solvent extracts. Most solvent extractions using CO_2 are run at temperatures between 10°C and 50°C, mild temperatures which are not likely to degrade or volatize heat-sensitive aroma compounds. Typically, the aroma of an $SCFCO_2$ extract more closely resembles the aroma of the feedstock botanical than the aroma of a steam distilled extract. Thus, an $SCFCO_2$ extract has more of the character of an absolute than a conventional solvent extract.

CO_2 PHASE EQUILIBRIA + FLUID DYNAMICS

<u>CO_2 Phase Equilibria</u> Knowledge of CO_2 phase equilibria is important because by understanding and applying of CO_2 phase equilibria theory, an efficient experimental plan can be formulated which

reduces process development R&D expenses, yet also minimizes the risk of specifying sub-optimal processing conditions for the commercial-scale plant. A theoretical understanding of how pressure, temperature, and cosolvents affect both the solvating power (important for rate and yield of commercial process) and selectivity (important for concentration levels of key components in commercial products) of the CO_2 solvent will greatly enhance the probability that each successive experiment yields process design data that can be used to further optimize commercial-scale plant performance and economics.

The critical pressure is the equilibrium pressure at the critical temperature; both together are referred to as the critical point. Above the critical temperature and pressure, CO_2 exists as a supercritical fluid (SCF). Figure 1 shows the relationship between the physical state of carbon dioxide and the temperature and pressure conditions in the system.

Supercritical fluids are unique because they have some properties which are similar to a gas others to a liquid. The density and solvating power can approach that of a liquid. Viscosity and diffusivity, however, are much closer to the properties of a gas. These characteristics are primarily responsible for the extraordinary solvent properties of $SCFCO_2$ and other supercritical fluids.

The solubility of a solute in general is proportional to solvent density. Liquid carbon dioxide is a non-polar (non-ionizing) solvent, in many ways like hexane. For liquid CO_2 near critical conditions, density increases rapidly with lowering temperature, being strongly temperature dependent. Above the critical temperature, density is both pressure and temperature dependent, rising with increased pressure, though falling with rising temperature. [Figure 2]

The effect of temperature on solubility is somewhat complex because of two competing effects; one effect tends to increase solubility with increasing temperature while the other tends to decrease solubility. As temperature increases, vapor pressure of the solute increases which tends to increase solubility, concomitantly CO_2 density decreases which tends to decrease solubility. For napthalene in CO_2 [Figure 3], for example, above 120 atm CO_2 density is less sensitive to temperature and therefore vapor pressure effects dominate. At 120 atm the two competing effects balance each other, and solubility remains relatively constant with increasing temperature. At pressures below 120 atm, $SCFCO_2$ density and therefore solubility are sensitive to small changes in temperature. The region where solubility decreases with increasing temperature, is called the "retrograde region".

Liquid carbon dioxide is a non-polar (non-ionizing) solvent, in many ways like hexane. However, in the supercritical phase the dielectric constant of CO_2 increases with increasing pressure.(3) The dielectric constant is an indicator of the polarity of the $SCFCO_2$ solvent [Figure 4]. Thus, by controlling pressure (or the addition of a polar cosolvent), the selectivity of SCF can be 'fine tuned' for the preferential extraction of the compounds of interest based on their polarity.

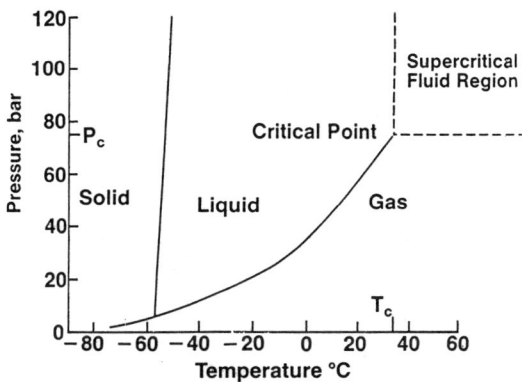

Figure 1. Carbon dioxide phase diagram.

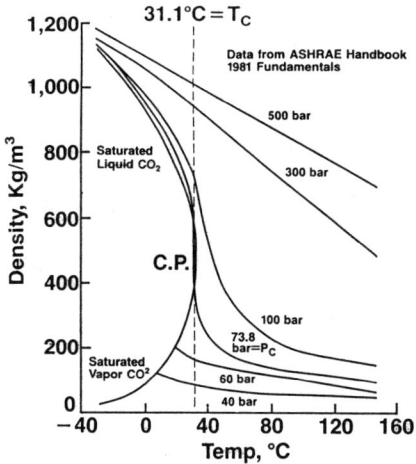

Figure 2. CO_2 density vs. temperature.

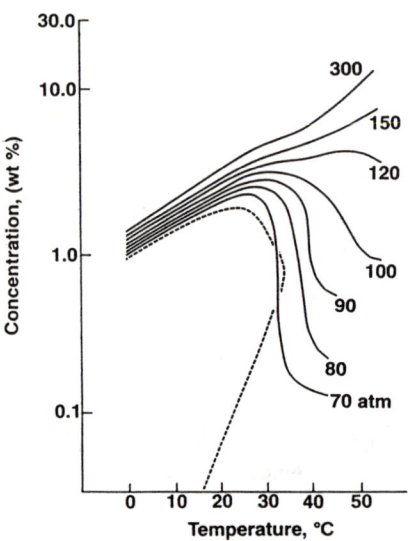

Figure 3. Naphthalene solubility in CO_2 vs. temperature. (Reproduced with permission from Reference 11. Copyright 1986 Butterworth.)

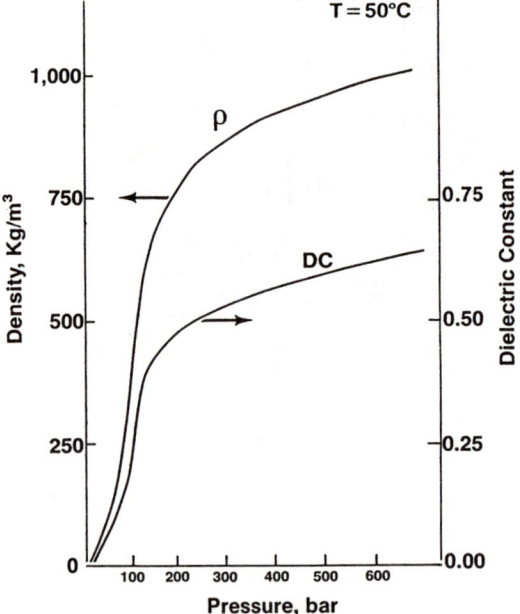

Figure 4. CO_2 density and dielectric constant vs. pressure. (Reproduced with permission from Reference 3. Copyright 1978 Verlag Chemie GmBH.)

Knowledge of phase equilibria theory is important to insure that screening experiments are structured to acquire data on phase equilibria properties that can significantly effect the commercial plant's performance. For example, in some cases a commercial-scale plant's performance and economic efficiency will be very sensitive to the introduction of a cosolvent (such as water and methanol) to the $SCFCO_2$ solvent. Experiments should be designed, utilizing knowledge of how addition of a cosolvent affects solubility of a compound of interest to yield data which can be used to determine if the addition of an appropriate cosolvent will significantly increase the commercial plant's performance and economic efficiency.

Fluid Dynamics Knowledge of fluid dynamics is important to insure that screening experiments are structured to acquire data on how fluid dynamic properties (such as $SCFCO_2$ solvent diffusivity and superficial velocity) will affect the commercial plant's performance and economic efficiency. For example, in many cases a commercial plant's performance and economic efficiency are sensitive to the superficial velocity of the $SCFCO_2$ solvent. Experiments should be designed to yield data that can be used to determine how sensitive commercial plant performance and economic efficiency is to superficial velocity of the $SCFCO_2$.

Solvent extraction from botanical substances can be characterized by four mass transport steps:
1) Diffusion of solvent into the botanical substrate.
2) Solvation of solute.
3) Diffusion of solute into the bulk fluid phase.
4) Transport of solute and the bulk fluid phase from the extraction zone.

Supercritical fluid extractions are typically run under conditions of high solvent/feed ratio, high superficial velocity, and low fluid viscosity. Thus, the controlling mass transfer parameter is usually the diffusion rate of the solvent and solute through the botanical substrate into the bulk fluid phase. Therefore, mass transfer rate can be increased by increasing solvent diffusivity, reducing diffusion distance, or elimination of diffusion barriers.

A far too common practice is to utilize CO_2 phase equilibria and fluid dynamics theory, but to limit experiments to the pressure, temperature, flow rate, cosolvent capabilities, etc. of the laboratory experimental apparatus that is conveniently at hand. Many commercial plant operating design inefficiencies could be eliminated by considering the effect on commercial plant performance and economic efficiency at the inception of the experimental program. Another caveat is that dramatic cost breaks occur in key commercial plant component equipment such as vessels, valves, heat exchangers, and pumps which are standard off-the-shelf commercial items at a given set of pressure, temperature and size specifications, but become a special order item at a marginally different set of operating conditions. Choosing experimental conditions that are cost effective for key commercial plant equipment requires the input of $SCFCO_2$ commercial plant design engineers at the early conceptual and R&D stages of process development to help ensure that the experimental program is targeted towards optimizing commercial plant performance and economics.

BOTANICAL STRUCTURE AND CHEMISTRY

Knowledge of botanical structure and chemistry is important for two reasons: 1) to determine if a pretreatment such as grinding, flaking, or rolling for particle size reduction and/or cell wall rupture will improve extraction efficiency, and 2) to determine the extraction conditions which maximize extraction rates, concentrations, and yields of value-determining components in the extract as well as to minimize the extraction rates, concentrations and yields of undesirable components in the extract product.

<u>Botanical Structure</u> Figure 5 illustrates the effect of flaking thickness on oil yield and rate of recovery from soybean feedstock. Rate of extraction increases and yield improves as flake thickness decreases. An oil yield of 97.4% was achieved from 0.10mm flakes; however, yield decreased to 87% and 67% for flake thicknesses of 0.38mm and 0.81mm respectively. (6)

Flaking reduces particle size, and hence, diffusion distance of the solute through the meal substrate. More importantly, as thickness decreases proportionately more cell walls are ruptured eliminating diffusion barriers. If cell walls encapsulate the valued components in a botanical substrate, pretreatment to rupture the cell walls significantly improves both yield and rate of extraction.

<u>Botanical Chemistry</u> Most botanical feedstocks contain a wide spectrum of compounds that can be dissolved in $SCFCO_2$. The character of the extract obtained from a botanical feedstock will markedly differ depending on the extraction and separation conditions chosen for the commercial-scale plant. Typically, as the $SCFCO_2$ density increases with increasing pressure, the solubility of less volatile components in the $SCFCO_2$ generally increases. Hence, by selecting suitable combinations of pressure and temperature for extraction and separation conditions, both selectivity and rate can be optimized for a commercial-scale plant design. For example, Figure 6 is a gas chromatogram of the chemical components found in a typical natural product. The desired extract product consists of all the essential oils and only the top fraction of waxes and resins; the extraction was accomplished at 4350 psi and 60°C. Upon complete extraction of the available essential oils, only the top fractions of the resins and waxes will have been extracted with virtually no pigments. Further fractionation was accomplished by a multi-stage separation process: a first stage to obtain a fraction rich in the top fractions of waxes and resins; a second stage rich in terpenes, free fatty acids and fatty oils; and a third and final stage to obtain a fraction rich in essential oils and esters. The product may be standardized by blending the first and third fractions, i.e. the resinoid and essential oil fractions. Thus, a high quality extract consisting of essential oil and soluble resins can be obtained having a remarkable resemblance to an absolute.

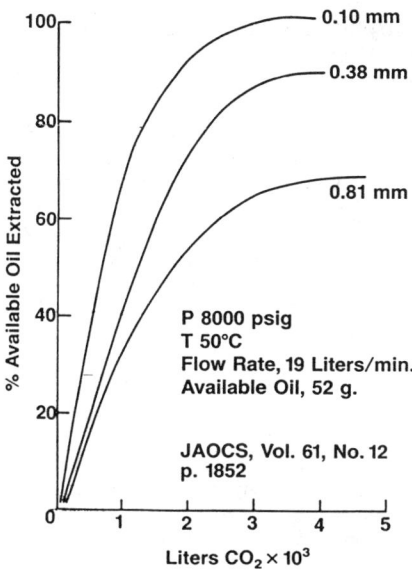

Figure 5. Soy flake thickness vs. extraction efficiency. (Reproduced with permission from Reference 6. Copyright 1984 American Oil Chemists' Society.)

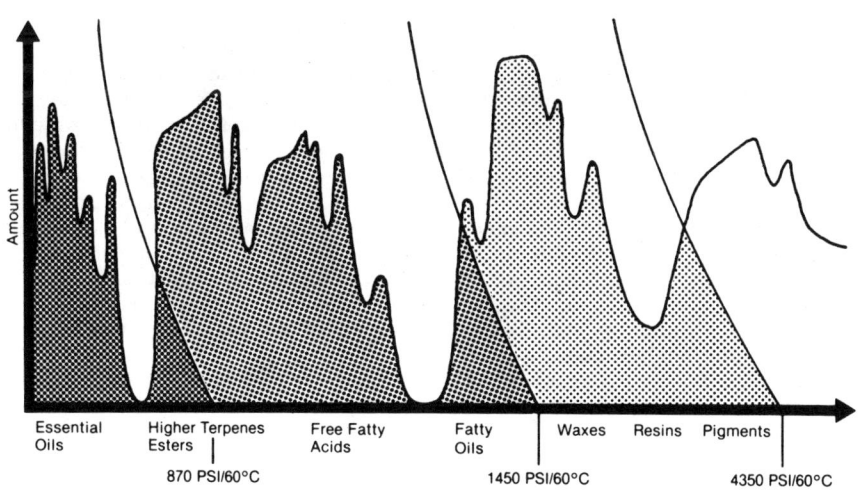

Figure 6. Gas chromatograms of a typical natural product. (Figure courtesy of Milton Roy.)

In general, $SCFCO_2$ selectively extracts essential oils, esters, alcohols, aldehydes, ketones, and lighter fractions of waxes and resins, leaving the less soluble heavier fractions of waxes and resins, fatty acids, triglycerides, chlorophylls, pigments and other high molecular weight species. Compared to $SCFCO_2$, conventional organic solvent extraction is much less selective. Sugars, many acids, starches, proteins and mineral salts are virtually insoluble in carbon dioxide. (2-5-10) In more complex situations, consideration must also be given to the way the component is bound to the botanical matrix. For example, alkaloids such as nicotine and caffeine are sometimes bound as complex salt with other compounds such as chlorogenic acid, citric acid, etc. and are difficult to extract without introducing water as a cosolvent to facilitate the extraction (3).

PROCESS DESIGN PROTOCOL

Yield and composition of an extract from a commercial-scale $SCFCO_2$ plant are determined by three key processing conditions:
1) Preparation of feedstock
2) Extraction conditions
3) Separation conditions

These parameters should be optimized for each product to be processed in the commercial-scale plant. However, in the case of a multiple product plant, compromises may be made and some products will be run at sub-optimal conditions to reduce capital and operating cost of the commercial-scale plant.

The process design protocol is a series of three successive levels of testing which further define the three key processing conditions with increasingly better resolution. Screening tests are quick and inexpensive and are mainly used to qualitatively define the processing sequence and determine the operating conditions for the more accurate process development unit (PDU) testing. If the screening unit tests show promise, then PDU testing would be initiated to optimize processing variables to generate data for economic evaluation of process alternatives that apply to the commercial-scale plant. Next, if economic evaluation of the process falls within the profitability goals of the commercial-scale plant, pilot plant testing is initiated to minimize scale-up uncertainty.

Screening Unit Testing (Assessing Technical Feasibility) The primary goal of screening unit testing is to assess technical feasibility. Experimental work begins using a supercritical fluid "screening unit" such as the Milton Roy system shown in Figure 7. The screening unit typically has one extraction vessel of 60-300 cc capacity and one or two separators.

Data from the screening unit is used to assess the process feasibility and product quality, and define the following process parameters:

Preparation of Feedstock	Extractor Conditions	Separator Conditions
Grating	Pressure	Pressure
Grinding (cryogrinding)	Temperature	Temperature
Rolling	Solvent/Feed Ratio	
Rapid depressurizing	Superficial Velocity	
Wetting		
Drying		

The extract product is analyzed to determine how changes in these parameters change extract yield and/or concentration. If the results of technical feasibility screening are encouraging, process development proceeds to the next step.

Process Development Unit (PDU) Testing (Optimizing Process Performance and Economics) The primary goal of PDU testing is to provide additional data for the key operating and process design variables to optimize process performance and economics. A PDU differs from a screening unit in several significant ways [Figure 8]. It has a) a larger extractor volume by about two orders of magnitude (10-20 liters vs. 60-300 cc); b) a CO_2 recovery system to study the effects of CO_2 recirculation; c) multiple separator vessels connected in series to study fractional separation of components; and d) optionally, a computerized data acquisition and control system to facilitate its operation and collection of data.

Data from the PDU experiments establish process conditions for preliminary process design. These parameters are listed below:

Extraction	Separation
Pressure	Pressure
Temperature	Temperature
Solvent/Feed Ratio	Fractional Separation
Superficial Velocity	Adsorbent Separation
Recycle Effects	Column Separation
Entrainers and Cosolvents	
Extractor Size, Shape and Configuration	
Processing Modes (e.g. Sub/Supercritical CO_2 or Dry/Wet CO_2 Extraction)	

Pilot Plant Testing (Scale-up Verification) The primary goal of pilot plant testing [Figure 9] is to minimize scale-up uncertainty. An $SCFCO_2$ pilot plant is typically 50-200 liters in extraction capacity. Pilot plant tests are designed to verify:
1) Optimum vessel configuration and size.
2) Effects of feedstock particle size and pretreatment techniques.
3) Mechanical design problems, such as clogging or salting-out of extract in piping, valves and heat exchangers; material handling considerations and cleaning requirements.

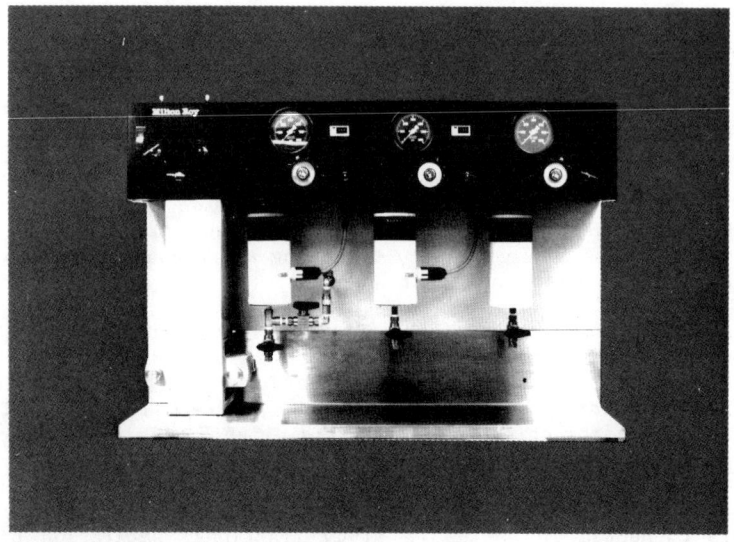

Figure 7. Milton Roy SCFCO$_2$ screening unit. (Photo courtesy of Milton Roy.)

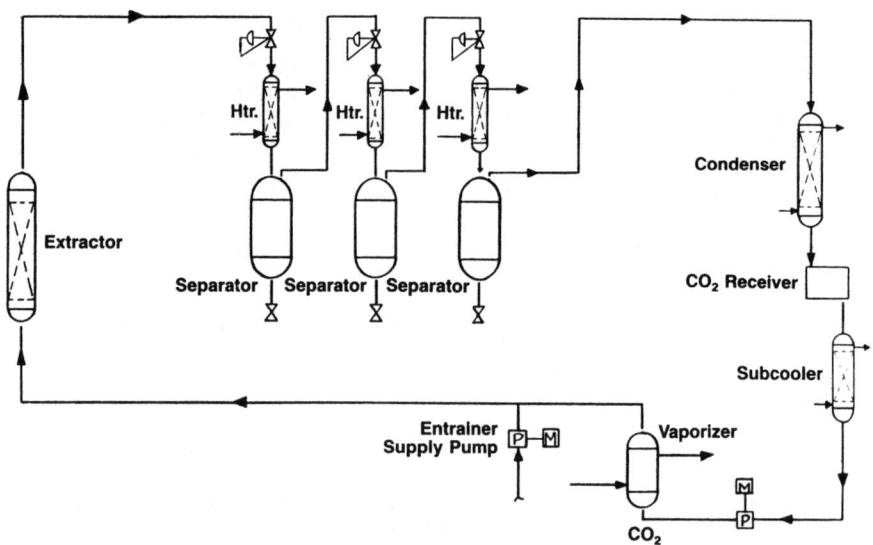

Figure 8. PDM process development unit schematic diagram.

COMMERCIAL PLANT DESIGN CONSIDERATIONS

Optimal commercial plant performance and economic efficiency requires investigation of several design alternatives to choose the most suitable commercial plant design. These commercial plant design considerations fall into two general categories - process design and mechanical design.

Process Design and Economic Evaluations

<u>Analysis of process development data</u> - The following illustrates how PDU data can be used to develop a commercial $SCFCO_2$ plant's process design. This data is provided by Marc Sims on pyrethrins, a natural insecticide extracted with subcritical and supercritical carbon dioxide from pyrethrum flowers (a species of chrysanthemum).

Figure 10 shows the effect of temperature on two parameters: pyrethrin recovery from the botanical substrate and concentration of the pryrethrin in the extract product. Pyrethrin yield rises with temperature reaching a maximum between 20-30°C but declines near 30°C. Selectivity for pyrethrin is lowest at 20°C and rises through and beyond the critical point.

Design engineers must utilize knowledge of the finished product requirements to determine the temperature of extraction. The minimum concentration of pyrethrin acceptable will determine extraction temperature. If 20% is acceptable, the extraction should be run at about 30°C. If a higher concentration is required, fractional separation, or a different operating pressure should be considered to enhance the concentration of pyrethrin in the product.

Figure 11 displays extraction efficiency vs. time on-stream. After two hours, 88% of the pyrethrin is recovered. Additional time marginally increases pyrethrin yield, however, total extract yield increases significantly diluting the pyrethrin. Since the additional components add no value to the extract, additional time is detrimental [Figure 12]. An economic evaluation balancing variables of pyrethrin yield and concentration against cycle time will determine the optimal time on-stream for this application.

<u>Alternative Fractionation and Standardization Techniques</u> - Raw material quality variations with respect to essential oils and other key ingredients require specialized processing schemes to produce extracts of standardized quality. The following are three common fractionation and standardization techniques that are indicative of the process design alternatives currently being used to achieve similar end product results. Economic evaluation of each alternative determines the best process design for a specific application.

The first example is the application of two-stage fractional extraction to selectively remove components from fennel (7). The first extraction stage operates under subcritical temperature conditions for extraction, i.e. 130 BAR and 30° C. Approximately 9% of the feed material is removed and the resulting extract is over 50% essential oil. The residue in the extractor vessel is then extracted a second time using supercritical conditions, i.e. 300 BAR

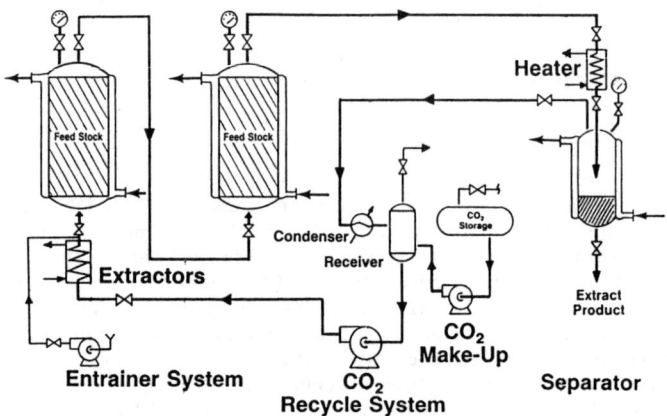

Figure 9. PDM pilot plant unit schematic diagram.

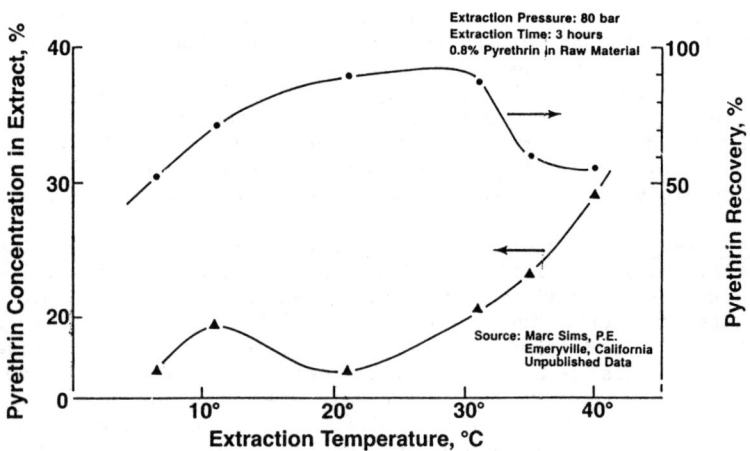

Figure 10. Efficiency of CO_2 extraction vs. temperature. (Printed with permission. Copyright 1987 Marc Sims.)

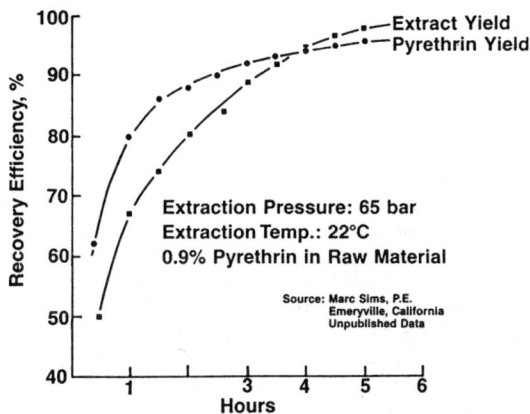

Figure 11. Efficiency of CO_2 extraction vs. time on stream. (Printed with permission. Copyright 1987 Marc Sims.)

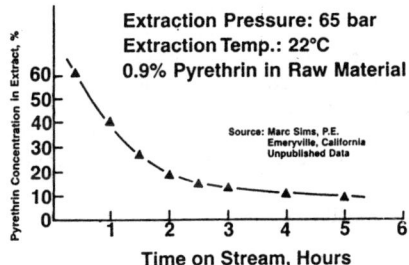

Figure 12. Pyrethrin concentration in extract vs. time on stream. (Printed with permission. Copyright 1987 Marc Sims.)

and 42°C. An additional 8.2% of the feed is extracted with very little essential oil in the extract. The two fractions can then be utilized separately or combined, as desired.

The second example (8) demonstrates two stage extraction using dry CO_2 followed by wet CO_2. Cinnamon is extracted at 300 BAR and 55°C. This first stage occurs with dry CO_2 to remove the essential oils responsible for the aroma and odor of the spice. The second extraction, using supercritical CO_2 saturated with water, extracts the flavor components. The fractions can be used separately or recombined to the desired composition.

A third example (9) uses fractional separation by a variation in separator temperature and pressure, usually with two or more separation vessels in series and by reducing the pressure in steps. This technique results in the separation of the less volatile components from more volatile compounds. Caraway seeds are extracted and precipitated in three stages, yielding three distinct fractions. [Figure 13] The first separator fraction contains only 1% essential oil and consists mainly of fatty oils. The second separator fraction, which amounts to only 9.6% of the extract, consists of a mixture of approximately one part essential oil to two parts fatty oils. The third separator fraction contains 90% caraway essential oil, reported to have excellent flavor characteristics.

Each method of fractionation requires different operating techniques and different equipment types, sizes and configurations. Two-stage extraction requires longer total extraction cycles, and possibly, parallel separator vessels to collect each fraction individually, and/or addition of humidification vessels. Multiple stage separation requires additional separator vessels with required heaters and pressure controls. In all three cases, the products can be used as separate fractions, or in various combinations to suit the user needs and marketplace considerations. Choice of fractionation and standardization methods for a commercial-scale $SCFCO_2$ plant is dependent on results of the economic evaluation of all technically feasible process design alternatives.

Mechanical Design, Construction and Start-up

Maintenance and Reliability - The selection of high pressure parts in rotating and/or reciprocating service must consider friction, lubrication, speed and similar factors to minimize wear, leakage and premature failure. Routine maintenance that can be performed in place with a minimum of off-line time is advantageous; if it becomes necessary to disconnect the unit from the piping or motor or move it to the maintenance shop for repair, spare units may be required.

An oversized unit operating at a slower speed may extend the period between preventive maintenance work as well as minimizing breakdowns or failures. High reliability is necessary for economical operations since a plant failure in the middle of a cycle results in wasted feed material, possibly wasted products, loss of CO_2, etc.

Cleaning Between Runs and Products - Sanitation within the system is a constant concern so special design considerations such as piping

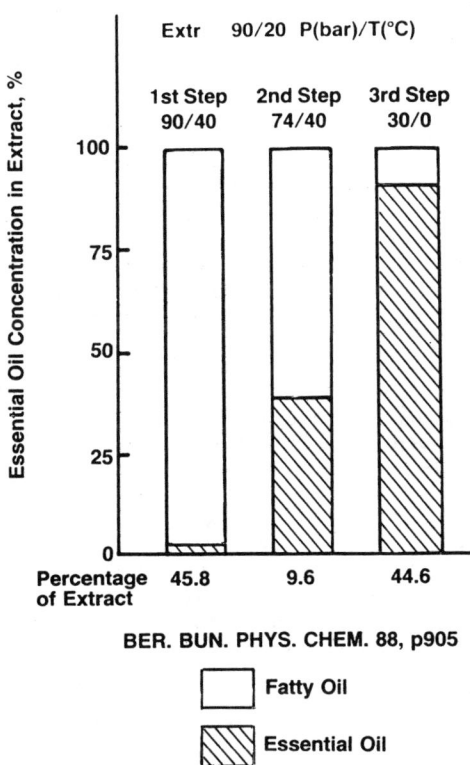

Figure 13. Caraway seed fractional separation. (Reproduced with permission from Reference 9. Copyright 1984 Verlag Chemie GmBH.)

disconnects at suitable points, elimination of dead legs where biologically active materials might accumulate, access to vessels and heat exchangers for cleaning, etc. In some cases clean-in-place (CIP) techniques will be sufficient.

Pretreatment and Posttreatment - Some materials may require pretreatment processing to prepare the material for extraction or post-treatment of the residue after extraction is completed. Material handling systems such as conveyors and product silos must be designed with these considerations in mind.

Piping and Valves - Some extracts develop a viscous oily phase; some extracts contain insoluble solids; some waxes or resins may precipitate within the piping, valves, or heat exchangers. The solvent CO_2 itself can deposit as a liquid phase or even (solid) dry ice if insufficient care is taken to determine the phase changes which may occur in piping and valving. Techniques to relieve plugging if it occurs must be designed into the system.

Seals and Gaskets - Seals, gaskets, and packings face a particularly harsh and unusual environment within the supercritical system. Elastomers with even low CO_2 absorption values may be unsuitable for the service because even a small volume of CO_2 absorbed into the elastomer at extraction pressure expands to many times its volume when pressure is reduced to atmospheric pressure. The rate at which pressure can be reduced without mechanically overstressing or destroying the elastomer must be determined. Areas where small quantities of CO_2 can be trapped at high pressure create problems when the pressure is relieved must be identified and eliminated. CO_2 is not a good lubricating fluid, so all lubricating oils and greases used must be FDA approved for food plant use.

Materials of Construction - Pure dry CO_2 is not particularly corrosive but in the presence of water, organic acids, etc. potential low cycle fatigue/corrosion problems exist. Selection of suitable materials of construction must be addressed during the plant design.

Quick-opening closures - Many of the vessels and heat exchangers within the system will operate in batch mode. Frequently it will be necessary to remove the top head of a vessel for introduction of the feed material or removal of the spent residue, or for cleaning and inspection of the surfaces in contact with the botanical. Several quick opening closure systems are currently available to accomplish this, but care must be taken to insure that the system chosen is best for the intended use.

Construction Schedule and Start-up - To date, the few commercial-scale $SCFCO_2$ plants constructed worldwide have, for the most part, experienced cost overruns due to retrofitting to correct design and construction flaws upon start-up, and the associated cost of start-up delay. A commercial-scale $SCFCO_2$ processing plant is highly capital intensive, and thus, profitability is very sensitive to timely construction and start-up. To best manage the problems of construction delay, and to expedite the solution of design and construction flaws that surface upon start-up, a single vendor should be held accountable for the responsibilities of engineering design, construction and start-up. 'Turnkey' construction contracts will greatly enhance the probability that future commercial-scale $SCFCO_2$ plants will be built on time and within budget.

SUMMARY

Developing a commercial supercritical CO_2 processing plant requires the application of five steps:
1) Application of high pressure CO_2 phase equilibria and fluid dynamics theory;
2) Application of knowledge of the botanicals structure and chemistry;
3) Performance of the process design protocol including screening studies, PDU studies, and pilot plant studies;
4) Utilization of process development data for process design and economic evaluation of alternative commercial plant designs;
5) Engineering, construction and start-up of the commercial plant.

ACKNOWLEDGMENTS

The Author wishes to gratefully acknowledge Mr. Marc Sims P. E., for his generous contribution of experimental process development data; Mr. Sam Vance P.E., Senior Process Engineer - PDM, Inc. for his many valuable contributions to both structure and content; and Pat Muska of PDM, Inc. for her efforts in typing this manuscript.

LITERATURE CITED

1) Taylor, F.M., Carbon Dioxide - The Solvent for the Food Related Industries, Wolviston Consultancy Services Limited, (Lichfield, England), 1984 pp. 29-31.
2) Moyler, David, Prepared Foods, November 1985, pp 100-101.
3) Hubert P., Vitzthum, O.G. Angew. Che. Int. Ed Engl. 17, p. 711, 1978
4) Hoyer G. G., Chemtech 1985, July, P. 471
5) Shultz, W.G., Schultz, T.H., Carlson, R.A., and Hudson, J.S., Food Technology, June, p. 32, 1974
6) Snyder, J.M., Frederick, J.P., and Christianson, D.D., JAOCS, 61 (12) pp. (1984).
7) U.S. Patent 4, 790,398
8) U.S. Patent 4, 198, 432
9) Stahl, E., Quirin, K.W., Glatz, A., Girard, D., and Riau, G., Busenges. Phys. Chem., 1984, 88, 900-907.
10) Rizvi, S.S., Daniels, J.A., Benando, Al.L., Zollweg, J.A., Food Technology, July, p. 59 1986.
11) McHugh, M. A. and Krukonis, V. J., Supercritical Fluid Extraction; Butterworth Publishers: Stoneham, MA 1986 p. 20

RECEIVED January 29, 1988

Chapter 8

Capillary Supercritical Fluid Chromatography with Applications in the Food Industry

T. L. Chester, L. J. Burkes, T. E. Delaney, D. P. Innis, G. D. Owens, and J. D. Pinkston

Miami Valley Laboratories, The Procter & Gamble Company, Cincinnati, OH 45239-8707

> A thermodynamic basis positioning the capabilities of
> SFC and GC is given and demonstrated with food
> applications. SFC will elute the same eluate at a
> lower temperature than GC, and will elute much higher
> molecular weight eluates than GC. This elution
> capability, when combined with universal detection via
> flame ionization, makes SFC a powerful technique for
> screening analytical problems within its scope. Mild
> temperatures, extended molecular weight range (compared
> to GC), and universal detection for organic compounds
> provide more accurate screening of samples with SFC
> than is possible with GC or HPLC.

Capillary supercritical fluid chromatography (SFC) ([1](#)) has characteristics fitting between those of conventional gas chromatography (GC) and high performance liquid chromatography (HPLC) ([2](#)). Thinking of SFC as an extension of GC, the supercritical mobile phase, with its characteristic liquid-like ability to solvate other materials, allows lower elution temperatures and a much larger molecular weight or volatility range than is possible in GC. But, supercritical fluids have much lower solute diffusion coefficients than gases. This generally results in longer analysis times for SFC than for GC. As an extension of HPLC, SFC is a faster technique when comparing liquid and supercritical mobile phases on the same column. Once again, this is because of differences in solute diffusion coefficients in the mobile phases. Thus, capillary SFC has reasonable analysis times while capillary HPLC usually takes too long to be a practical problem-solving tool. What is even more important in considering practicality and problem-solving potential is the compatibility of many supercritical fluid mobile phases with GC detectors ([3-8](#)) and with spectroscopic detectors ([9,10](#)). In particular, the use of CO_2 with the flame-ionization detector combines separations (with a low-temperature, solvating mobile phase of continuously adjustable and programmable strength) over a greatly extended molecular weight range (compared to GC) with universal detection for organic compounds.

0097-6156/88/0366-0144$06.00/0
© 1988 American Chemical Society

Understanding why SFC elution behaves in this manner is most easily achieved by starting from a GC perspective. If we study the retention behavior of a single, stable analyte on a capillary GC column as a function of temperature, we find that the logarithm of the capacity ratio varies linearly with the reciprocal of the absolute temperature. The capacity ratio, k, is $(t_r-t_o)/t_o$ for isothermal conditions. The variables t_r and t_o are the retention times of the analyte and an unretained material, respectively. This GC retention behavior is shown in Figure 1.

A thermodynamic explanation of GC retention vs. temperature is based on the Van't Hoff equation, extended for GC:

$$\ln k = -\Delta G_s/RT - \ln \beta \qquad (1)$$

where ΔG_s is the free energy of solution of the solute in the stationary phase. R and T are the gas constant and absolute temperature, respectively, and β is the phase ratio of the column (defined as the ratio of the mobile phase to the stationary phase volumes). Upon substituting $\Delta H_s - T\Delta S_s$ for ΔG_s, we see the slope of the retention curve gives the enthalpy of solution of solute in the stationary phase, ΔH_s, and the intercept is determined by the phase ratio and the entropy (ΔS_s) terms. This retention behavior for a given analyte is a function only of the column stationary phase, phase ratio and the temperature. Retention is completely independent of the mobile phase as long as it is inert and non-solvating under the conditions of the measurements. Thus, He, H_2, and even CO_2 produce the same retention behavior under GC conditions.

Now, suppose we partially plug the column outlet while raising the inlet pressure. With a CO_2 mobile phase we can achieve significant solvating densities beginning at pressures of about 70 atmospheres. We have to choose and adjust the restriction at the outlet to get the mobile phase velocity range we want for the operating pressure range. As mobile phase solvation begins, the mobile phase competes with stationary phase for solute. The net effect is a reduction in analyte retention. In our thermodynamic model, a new term must be added representing the effect of mobile phase solvation:

$$\ln k = -\Delta G_s/RT - \ln \beta + \Delta G_m/RT \qquad (2)$$

or

$$\ln k = -(\Delta G_s - \Delta G_m)/RT - \ln \beta \qquad (3).$$

The new term, ΔG_m, represents the free energy of solution of solute in the mobile phase. The magnitude of this term depends on the density of the mobile phase, and is zero for non-solvating mobile phases (as in GC). Upon substitution and simplification, equation 3 becomes

$$\ln k = -(\Delta H_s - \Delta H_m)/RT + C \qquad (4)$$

where C is the intercept containing the phase ratio and entropy terms. Looking at retention data again on a Van't Hoff plot, we see that increasing the mobile phase density decreases the slope and reduces retention (11). Mobile phase density has a large effect on

the enthalpy of solution and the slope of the plot. But if
stationary phase swelling is negligible (fixing the phase ratio) and
retention occurs purely by a partition mechanism (without
adsorption), then the intercept should be constant (11). This ideal
behavior is shown in Figure 2, a Van't Hoff plot of a single,
perfect analyte at several different mobile phase densities. In
actual fact, the intercept does move somewhat because the stationary
phase swells with increasing mobile phase density (12). But if the
stationary phase even doubles its volume, it causes only a
relatively small shift of the intercept because of the log function.

Even though the movement of the intercept has not been studied
in depth in capillary SFC, the general trend of retention curves at
various densities at temperatures away from the origin are well
known: Increasing the mobile phase density at any temperature
reduces the retention. Several conclusions follow from this
knowledge:

>> Any compound that can be eluted from a given column by GC
can be eluted with less retention by SFC.

>> Any compound that can be eluted from a column by GC can be
eluted with lower temperature by SFC.

>> For any homologous or oligomeric series of compounds, SFC
is capable of achieving a higher molecular weight range
than GC.

>> Any technological improvements that extend the scope of GC
will similarly extend the scope of SFC.

So, with SFC, temperatures as low as 35°C are possible, and a
molecular weight range of 3000 is fairly typical (although masses
exceeding 10,000 Da have been eluted and detected with an FID). And,
compared to HPLC, SFC gives us capillary-column efficiency combined
with detector options not possible with organic mobile phases.

The price we must pay for this capability comes in terms of
analysis time. If we compare all three types of mobile phase on one
column, we see the column efficiency behavior illustrated in Figure
3. Here we've simply evaluated the Golay equation (13) (giving
height equivalent to a theoretical plate, h) for liquid, supercritical fluid at two densities, and gas mobile phases, and plotted
the results as a function of the log of the mobile phase velocity.
The Golay equation predicts that any column will produce the same
minimum h regardless of the mobile phase used. The mobile phase
choice determines the optimum velocity (where the minimum h occurs)
and thus has a direct effect on the analysis time. This is caused
by the differences in the solute diffusion coefficients between the
mobile phases. As seen in the figure, GC is the fastest technique,
by far. SFC is second. And HPLC is slowest -- in fact, capillary
HPLC is so slow it is not widely used. But, in practice, one column
is not used for all three techniques. Short packed columns are used
successfully in HPLC and give very short analysis times. Packed-
column SFC can be even faster. But, without the use of (detector-
limiting) organic co-solvents, packed-column SFC has not yet matched
the molecular weight range of capillary SFC.

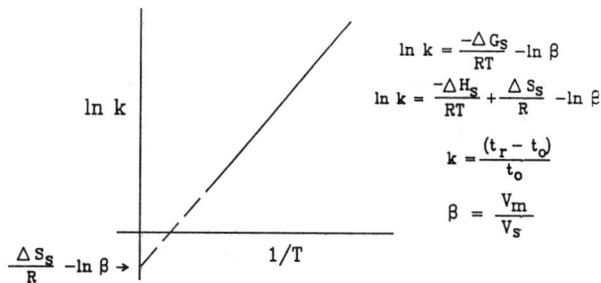

Figure 1. Retention behavior of an ideal analyte in gas chromatography.

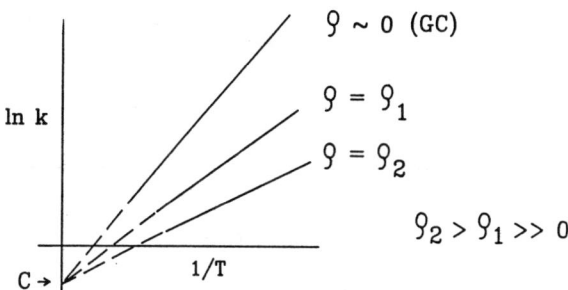

Figure 2. Retention behavior of an ideal analyte in supercritical fluid chromatography with changing mobile phase density, ρ.

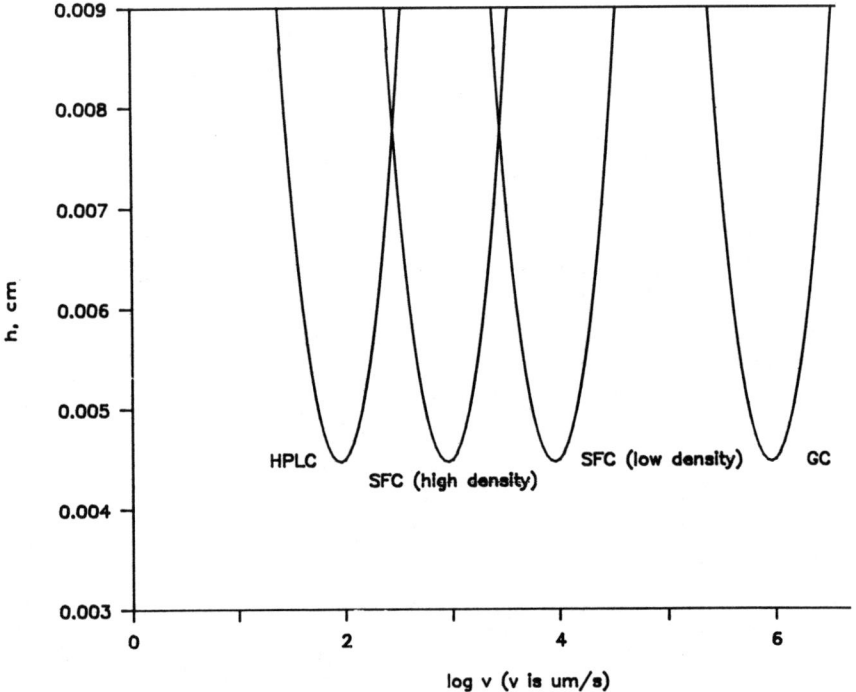

Figure 3. Calculated height equivalent to a theoretical plate, h, vs. log of the mobile phase velocity for a perfect analyte eluted on one capillary column with three different mobile phases. Assumptions are: column inside diameter is 50 um; diffusion coefficients are 10^{-1} cm^2/s for gas, 10^{-3} and 10^{-4} cm^2/s for low- and high-density supercritical fluid, respectively, and 10^{-5} cm^2/s for liquid.

Comparing only the viable capillary column techniques, GC and SFC, the columns used in each are not identical. Shorter, smaller-diameter columns are chosen for SFC to reduce the analysis time. The smaller diameters also lower (that is, they improve) the minimum plate height. However, column efficiency actually realized is more important than plate height. So, the column length and mobile phase velocity used must also be considered. Figure 4 compares efficiencies of typical SFC and GC columns as a function of mobile phase velocity. The figure is somewhat deceptive at first glance. The SFC column has a much higher optimum efficiency but is usually used at a higher-than-optimum velocity to shorten analysis time: Efficiency is traded for speed of analysis. The velocity and efficiency ranges commonly used in both techniques are indicated by the shaded areas. Thus, in common practice, capillary SFC has a lower efficiency and a somewhat longer analysis time than GC.

So, whenever it is capable of solving a problem, GC is the best choice by virtue of its speed and efficiency. But the limited molecular weight range and the uncertainties of analyte survival at high temperatures are significant restrictions in the use of GC. This is obvious in cases where known compounds fail to elute. However, these restrictions may be subtle, and potentially more costly, when the thermal stability and volatility of the sample components is unknown. HPLC is restricted by detector limitations -- detector choice is a guessing game when the analytes are unknown. SFC is superior to GC and HPLC for the first screening of many new samples. For this purpose SFC provides:

>> A lower risk of thermally degrading (and altering) the sample than GC.

>> A larger molecular weight range than GC, providing a more complete picture of what kinds of things are in the sample. (GC provides more low-molecular-weight detail, but that can come later, if required).

>> A more complete picture of what is present within its mass range (provided by the FID) than is possible by HPLC with its detector limitations.

SFC is extremely valuable for screening problems and for validating GC and HPLC methods when they are best. And, SFC proves to be the best technique for routine analyses in many cases.

Experimental Considerations

Capillary SFC requires a pulse-free, pressure-controlled pump, an injector, a column, a column oven, and a detector. The basic instrumentation has been described in a number of reports and reviews (14,15). In addition, with the flame-ionization detector, the mass spectrometer, and any other low-pressure detector, a flow-restricting interface is required between the column outlet and

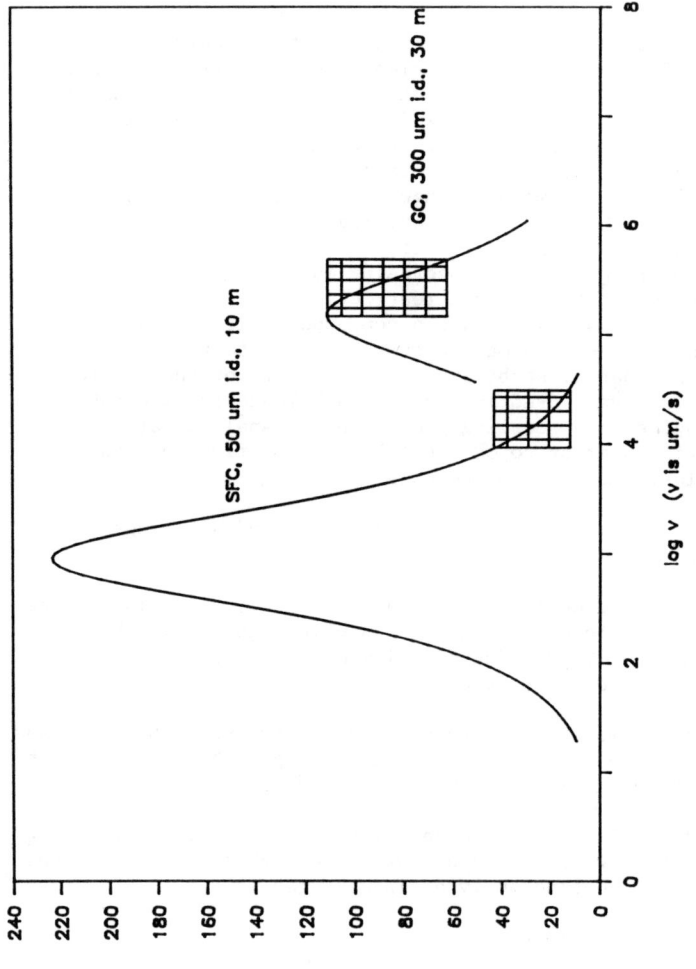

Figure 4. Column efficiencies of typical SFC and GC columns vs. log of the mobile phase velocity. Diffusion coefficients used are 10^{-1} and 10^{-4} cm^2/s for gas and supercritical fluid, respectively. The velocity and efficiency ranges commonly used for each technique are indicated by the shaded areas.

the detector. This is necessary to control the mobile phase
velocity while maintaining the mobile phase pressure (and density)
over the entire length of the column. This also has been the
subject of much previous work (3-7).

Columns are made of fused silica in a fashion similar to modern
GC columns. The dimensions are somewhat different than used in GC.
Inside diameters are usually 50-100 um, and the thickness of the
stationary phase film relative to the inside diameter is larger than
in GC (16). Stationary phases must be insoluble in the mobile
phase. Crosslinked silicones are very stable, especially with CO_2
mobile phase, and have been used as stationary phase in most of the
capillary SFC work reported to date.

Applications

We will focus on examples that are either impossible, or at least
questionable, by GC and HPLC.

SFC works great for a wide variety of oligomers. Ethoxylates
and propoxylates have many uses in the food industry. Yet, they are
difficult to characterize because of their mass range, often
exceeding several thousand daltons, combined with poor light
absorbance which limits HPLC detection. SFC is able to elute
polyethylene glycols out to about E_{45} (that is, an ethoxylate
chain with 45 members) as seen in Figure 5. The ability of SFC to
handle highly-polar glycols is important when glycol chains are
attached to nonpolar molecules to make surfactants and emulsifiers.

The superiority of SFC in analyzing oligomers is easily seen in
the analysis of silicones. Silicones (in this case,
polydimethylsiloxanes, or PDMS) are specified according to their
viscosity. A 20 cs PDMS can be separated over most of its molecular
weight range using GC. However, heavier, more viscous samples
cannot be differentiated by GC. HPLC, and especially gel permeation
chromatography (GPC), is frequently used to get molecular weight
distributions of PDMS samples. However, the chromatogram is usually
just an envelope with no information on individual components and no
indication if impurities exist over the same molecular weight range
as the sample. Figures 6 and 7 show SFC chromatograms of two 500 cs
PDMS samples, both of which met manufacturers' specifications. The
mass range in these chromatograms is estimated to extend beyond 8500
Da (17). Resolution of individual components gradually diminishes
until being completely lost about half way through the
chromatograms. This is due to the increasing similarity of adjacent
series members with increasing degree of polymerization and to the
increasing solute diffusion coefficients with increasing solute
molecular weight and mobile phase density. GPC chromatograms of
these two samples are very similar to each other and show no detail
of individual peaks. GC is pointless. Yet, SFC, while not perfect,
provides the most detailed information available, easily
distinguishes these samples from each other, and is capable of
detecting (early-eluting) overlapping series of impurities.

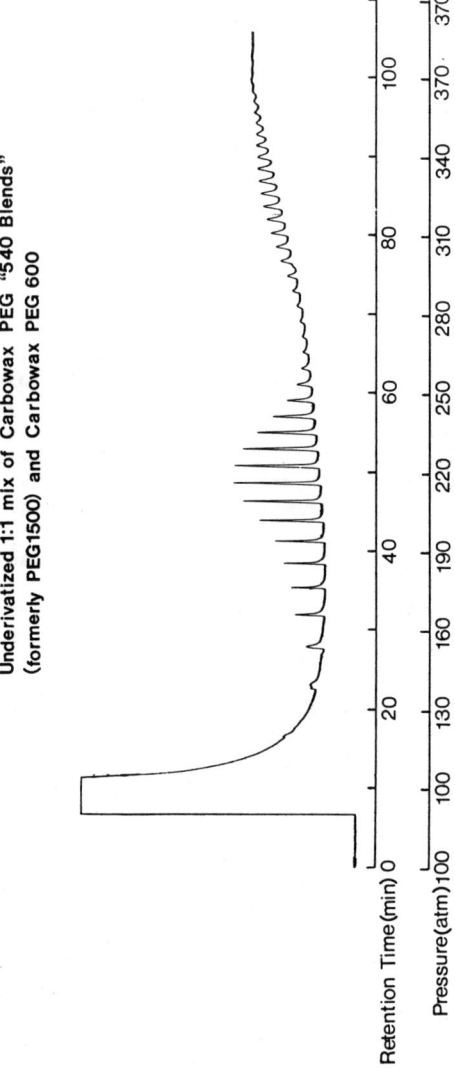

Figure 5. Chromatogram of an underivatized polyethylene glycol mixture, 2% w/w in methylene chloride. The injection volume was 0.10 ul with a 1:10 split. A 10 m x 50 um i.d. DB-17 column with a 0.1 um film thickness was used. The mobile phase was CO_2 used at an oven temperature of 100°C.

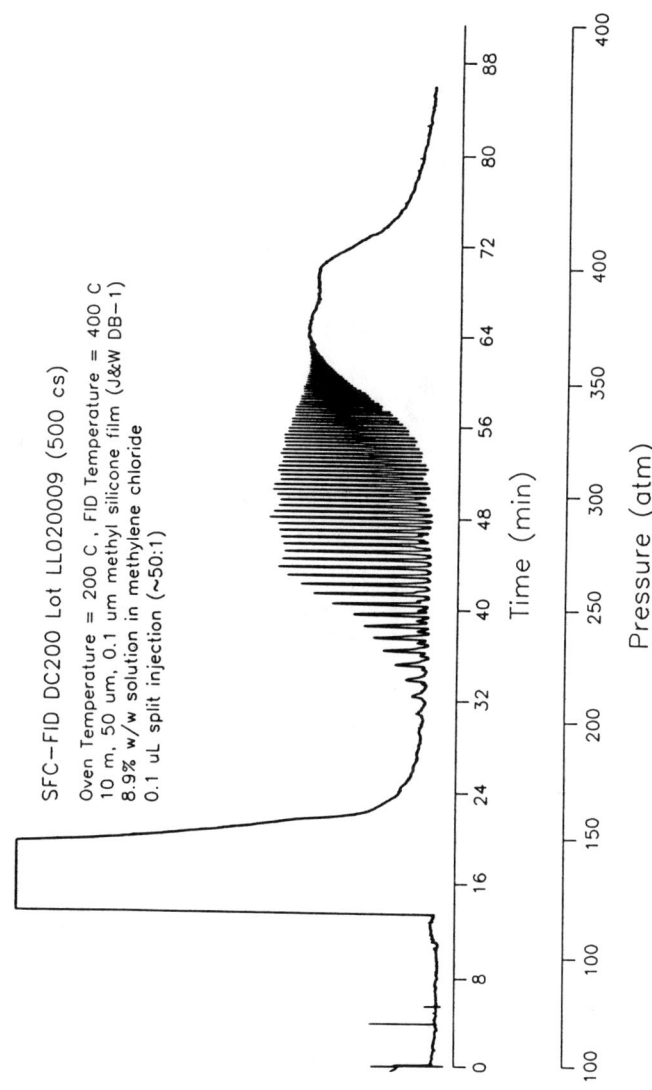

Figure 6. Chromatogram of a 500 cs polydimethylsiloxane sample.

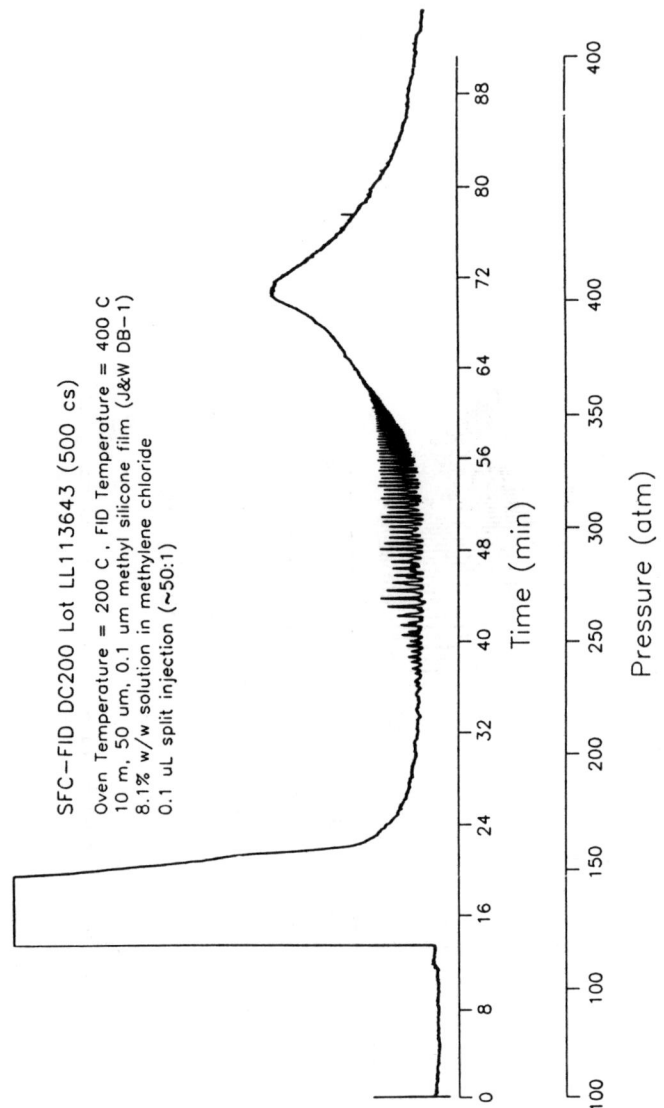

Figure 7. Chromatogram of a second 500 cs polydimethylsiloxane sample, supposedly identical to the sample in Figure 6.

The ability of SFC to elute silylated oligosaccharides through about G_{18} (that is, a chain of 18 glucose units) has been reported (18). A typical SFC-FID chromatogram of a silylated corn syrup solids sample is shown in Figure 8. The double peaks are believed to be caused by the two possible anomers at each molecular weight. Ordinarily, oligosaccharides cannot be dissolved well enough in CO_2 for a practical analysis. However, they silylate easily in just one step. The silylated analogs are very soluble in CO_2 and give excellent SFC behavior. Conventional GC is the best method to analyze oligosaccharides through about G_5, but requires temperatures exceeding 350°C to go any further. If columns could be developed with 500°C temperature ranges, and if the sample components could survive such a high temperature analysis, extrapolation of lower-molecular-weight retention data suggests that GC could only reach the G_{10}-G_{12} range. HPLC has more molecular weight range than has been demonstrated so far even for SFC, but can only resolve a few members of a series on any one injection due to the isocratic solvent limitation imposed by the detectors.

Earlier we emphasized how effective SFC-FID can be at surveying samples new to a particular lab. The SFC-FID combination gives a much better chance of seeing what is in the sample than either GC or HPLC used alone. The final examples were all done in survey fashion with minimum sample preparation and with no attempt to determine the identities of the peaks.

Emulsifiers and gums are important in the food industry and are often hard to characterize. Figure 9 shows chromatograms of a chewing gum extract and a bubble gum extract. Each was prepared by blending a stick of chewing gum or a piece of bubble gum in a 1:1 mixture of water and methylene chloride. The methylene chloride layer was then separated, filtered, and injected. The major peaks eluting above 250 atmospheres, especially in the cinnamon-flavored chewing gum, are noteworthy. Generally, peaks eluting at 250 atmospheres (at a 100°C column temperature) by capillary SFC would require an elution temperature of over 350°C using GC. And, significant additional scope is available with SFC since the pressure range available exceeds 400 atmospheres.

Figure 10 is a chromatogram of honeycomb extract. The honeycomb was simply dissolved in toluene, filtered, and injected. Honeycomb contains fatty acids, paraffins, esters, diesters, and other components. It is doubtful that all the peaks visible in Figure 10, and especially those eluting above 300 atmospheres, could be eluted by GC.

Figure 11 shows chromatograms of black and red pepper extracts. Here is a case where GC may be suitable for the black pepper but not for the red pepper because of the major, late-eluting peaks.

We have given theoretical and practical arguments and examples to illustrate how SFC can be used to solve separation problems when conventional techniques are not sufficient. SFC would be a powerful addition to the capabilities of a well-equipped analytical lab, and would be especially valuable to aid in the characterization of unknown samples.

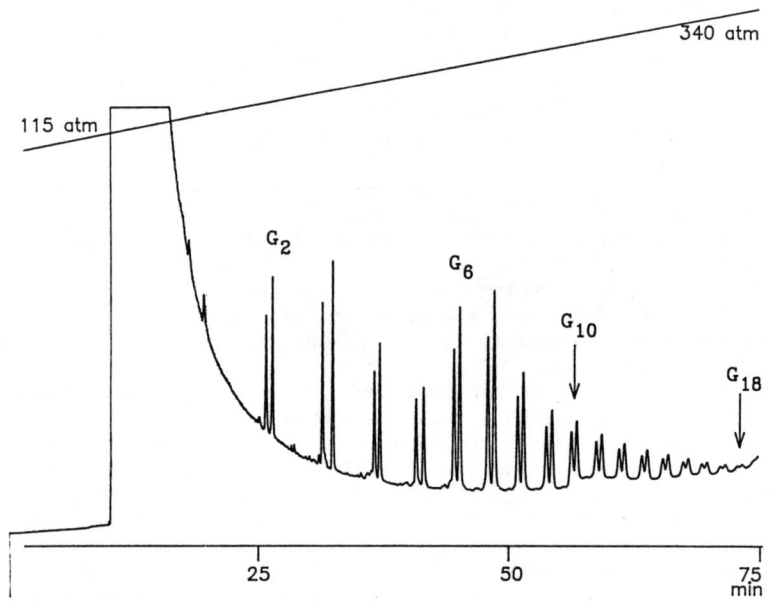

Figure 8. Chromatogram of silylated Maltrin 100 (corn syrup solids). Conditions: 10 m x 50 um i.d. DB-1 column, CO_2 mobile phase at 90°C, pressure programmed as indicated, FID at 400°C.

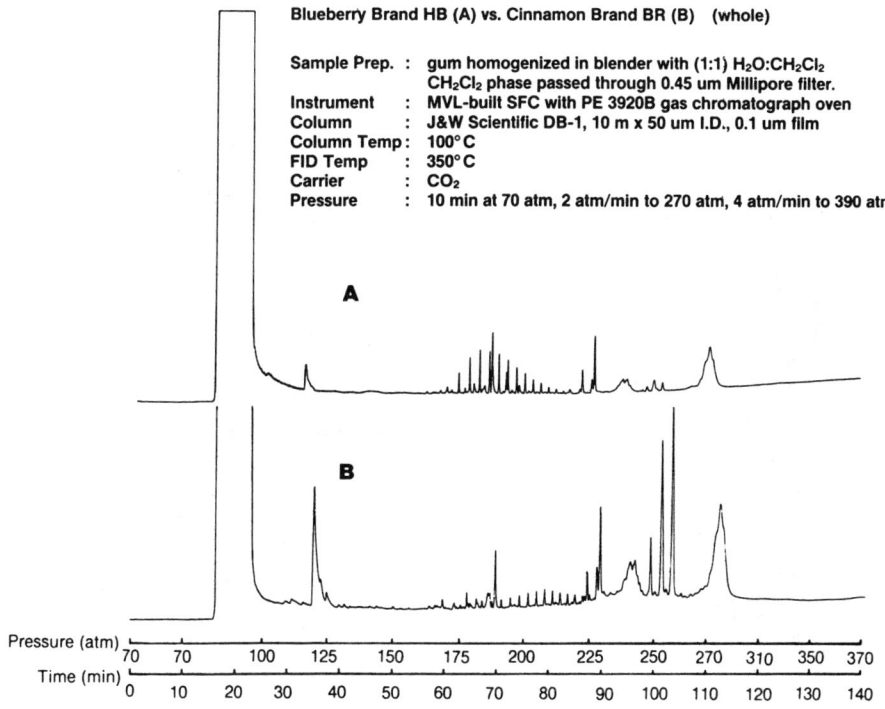

Figure 9. Chromatogram of extracts of two different chewing gums.

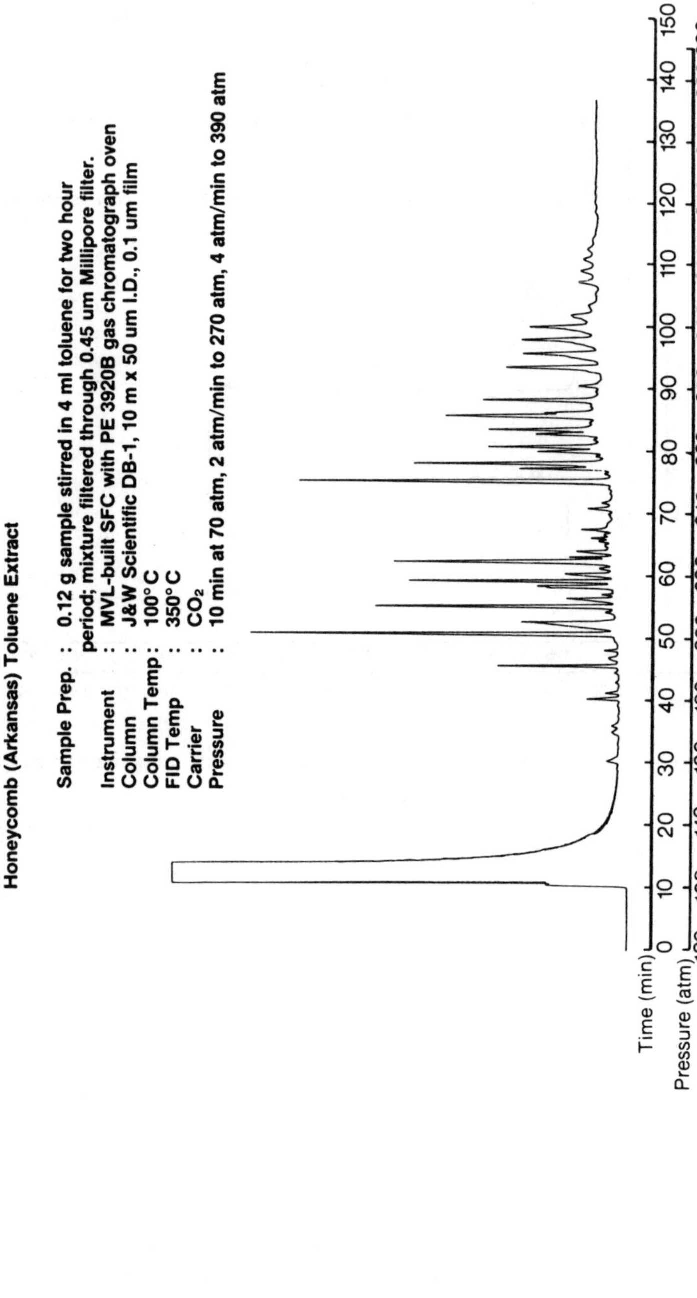

Figure 10. Chromatogram of honeycomb extract.

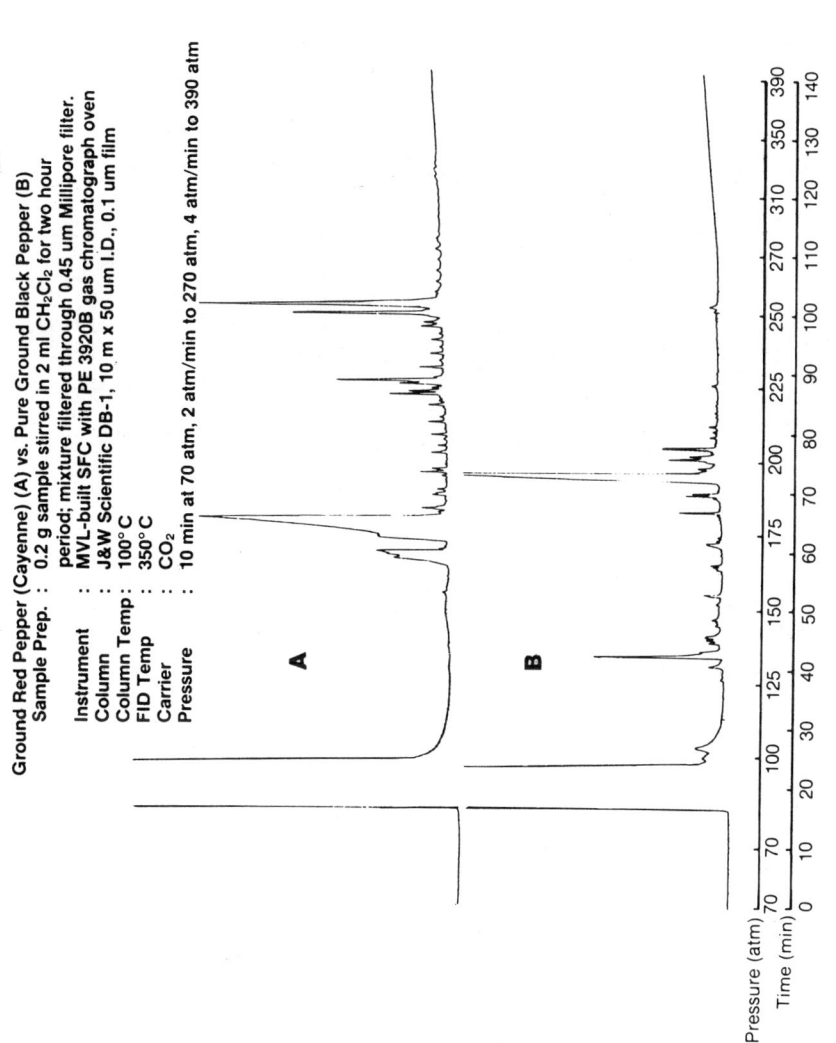

Figure 11. Chromatograms of extracts of red and black pepper.

Literature Cited

1. Novotny, M.; Springston, S. R.; Peadon, P. A.; Fjeldsted, J. C.; Lee, M. L. Anal. Chem. 1981, 53, 407A-414A.
2. Chester, T. L. J. Chromatogr. Sci. 1986, 24, 226-9.
3. Fjeldsted, J. C.; Kong, R. C.; Lee, M. L. J. Chromatogr., 1983, 279, 449-55.
4. Chester, T. L. J. Chromatogr. 1984, 299, 424-31.
5. Chester, T. L.; Innis, D. P.; Owens, G. D. Anal. Chem., 1985, 57, 2243-7.
6. Richter, B. E. J. High Resoln. Chromatogr./Chromatogr. Comm., 1985, 8, 297-300.
7. Guthrie, E. J.; Schwartz, H. E. J. Chromatogr. Sci., 1986, 24, 236-41.
8. Markides, K. E.; Lee, E. D.; Bolick, R.; Lee, M. L. Anal Chem. 1986, 58, 740-3.
9. Wright, B. W.; Kalinoski, H. T.; Udseth, H. R.; Smith, R. D. J. High Resoln. Chromatogr./Chromatogr. Comm., 1986, 9, 145-53.
10. Shafer, K. H.; Griffiths, P. R. Anal. Chem., 1983, 55, 1939-42.
11. Yonker, C. R.; Smith, R. D. J. Chromatogr., 1986, 351, 211-218.
12. Springston, S. R.; David, P.; Steger, J.; Novotny, M. Anal. Chem., 1986, 58, 997-1002.
13. Golay, M. J. E. In Gas Chromatography, Desty, D. H., Ed.; Butterworths, London 1958.
14. Fjeldsted, J. C.; Lee, M. L. Anal. Chem., 1984, 56, 619A.
15. Lee, M. L.; Markides, K. E. Science, 1987, 235, 1342-7.
16. Peadon, P. A.; Lee, M. L. J. Chromatogr., 1983, 259, 1-16.
17. Owens, G. D.; Burkes, L. J.; Pinkston, J. D.; Keough, T.; Simms, J. R.; Lacey, M. P., ACS Symposium Series, this volume.
18. Chester, T. L.; Innis, D. P. J. High Resoln. Chromatogr./Chromatogr. Comm., 1986, 9, 209-212.

RECEIVED November 13, 1987

Chapter 9

Retention Processes in Supercritical Fluid Chromatography

Clement R. Yonker and Richard D. Smith

Chemical Sciences Department, Pacific Northwest Laboratory, Richland, WA 99352

>   The understanding of retention processes in
> supercritical fluid chromatography is important for the
> continued growth and development of the analytical
> technique. Chromatographic retention has been studied
> using a simple thermodynamic model involving solute
> retention as a function of pressure and through
> assessment of the effect of density on the enthalpy of
> solute transfer between the supercritical mobile phase
> and bonded stationary phase. The solvatochromic behavior
> of solute probes in pure and binary supercritical fluids
> can be used to determine their polarity/polarizability as
> a function of density and related solvent effects on
> solute retention in supercritical fluid chromatography.

Supercritical fluid chromatography (SFC) is a rapidly expanding analytical technique for the separation of a wide range of compound classes, some of which are not easily amenable to gas chromatography or liquid chromatography. It can be argued that the continuation of this rapid growth and expansion into other analytical applications would benefit from an improved fundamental understanding of solute retention in SFC using both pure and binary supercritical fluids as mobile phase solvents. The basic physicochemical studies to be discussed in this chapter effect not only chromatographic retention, but also supercritical fluid extraction processes and chemical reaction mechanisms in supercritical fluids.(1-4)

   The distribution coefficient for the chromatographic retention process is similar to the bulk distribution coefficient studied in supercritical fluid extraction. Reaction rates in supercritical fluid solvents as a function of pressure are determined by the change in partial molar volume ($\Delta \bar{v}_i$) between the products and reactants.(4,5) The same parameter ($\Delta \bar{v}_i$) affects solute retention in SFC, as will be shown in the Theory section. Therefore, the study and understanding of solute retention in SFC is relevant to all technological applications of supercritical fluids.

   Three major areas of investigation in our laboratory using supercritical fluids will be discussed. These entail: 1) solute retention modeling using a simple thermodynamic description of retention as a function of pressure at constant temperature in SFC;

0097-6156/88/0366-0161$06.00/0
© 1988 American Chemical Society

2) the investigation of the thermodynamics of solute retention by studying the behavior of the enthalpy of solute transfer between the fluid mobile phase and the bonded stationary phase, for both pure and binary supercritical fluids, and 3) the study of the solvatochromic behavior of pure and binary supercritical fluids at different pressures and temperatures. The latter method allows one to study the solvation process (solute-solvent interactions) in the fluid mobile phase as a function of density independent of the presence of the bonded stationary phase.

The combination of these fundamental studies provides a basis for an integrated approach to understanding retention processes in SFC. Inherent in this approach is a concommitant increase in the understanding of supercritical fluids for applications in extraction and chemical reaction phenomena. In this chapter the results of our initial studies on solute retention in SFC will be presented and discussed.

Theory

Thermodynamics of Retention in SFC. Solute retention in SFC as a function of pressure is given as, (6,7)

$$(\partial \ln k'/\partial P)_T = 1/RT[\bar{v}_1^{mp,\infty} - \bar{v}_1^{sp,\infty}] - K \tag{1}$$

where k' is the dimensionless retention factor of the solute, R is the gas constant, T is temperature, $\bar{v}_1^{mp,\infty}$ and $\bar{v}_1^{sp,\infty}$ are the solute partial molar volumes in the mobile and stationary phases at infinite dilution respectively, and K, the isothermal compressibility of the fluid solution, is defined as $1/V (\partial V/\partial P)$. This term relates the change in molar volume of the compressible fluid (V) as a function of pressure to solute retention in SFC. Under the experimental conditions discussed in this chapter (and almost universally relevant to SFC), the mole fraction of the selected solute, naphthalene, in the fluid phase is very low ($X_1^{mp} << 10^{-2}$) (8), therefore, $\bar{v}_1^{mp} \sim \bar{v}_1^{mp,\infty}$.

The partial molar volume of the solute in the mobile phase at infinite dilution is a sensitive probe of the solvent-solute interactions. The partial molar volume of the solute at infinite dilution can be expressed as two terms through a triple product relationship.

$$\bar{v}_1^{mp,\infty} = K_2 V_2 \ [(\partial P/\partial n_1)_{T,V,n_2}] \tag{2}$$

where K_2 is the isothermal compressibility of the pure solvent and V_2 is the molar volume of the pure solvent. (9,10) Substitution of equation 2 into equation 1 yields a relationship for the slope of solute retention in SFC.

$$(\partial \ln k'/\partial P)_T = 1/RT\{K_2 V_2 (\partial P/\partial n_1)_{T,V,n_2} - \bar{v}_1^{sp,\infty}] - K \tag{3}$$

From equation 3, using an appropriate equation of state (EOS) that describes the P,V,T equilibria for the solute and the fluid, solute retention in SFC can be predicted for systems falling within the scope of our assumptions.

The density dependence of retention at constant temperature is defined as,

$$(\partial \ln k'/\partial \rho)_T = (\partial \ln k'/\partial P)_T \, (\partial P/\partial \rho)_T \tag{4}$$

where ρ is density (g/cm^3). The first term on the right hand side (RHS) of equation 4 has already been evaluated in the previous equation. The second term on the RHS of equation 4 is the slope of pressure versus density at constant temperature for the fluid and can be readily obtained from an EOS. The Peng-Robinson, two-parameter, cubic EOS was used to evaluate the partial molar volume of the solute in the mobile phase (equation 2), the isothermal compressibility of the fluid solution and solvent, and $(\partial P/\partial \rho)_T$ in Equation 4.

For systems composed of a pure fluid solvent and an organic solute, standard mixing rules were used for the calculation of the mixture parameters (10,11). Therefore, all the terms in equation 3 and 4 can be evaluated as a function of pressure and density at constant temperature, except for the partial molar volume of the solute in the stationary phase at infinite dilution $\bar{v}_1^{sp,\infty}$. The partial molar volume of the solute in the stationary phase can be estimated, allowing solute retention in the supercritical region to be calculated as a function of pressure.

Assuming conditions of infinite dilution during retention in SFC, the solute distribution coefficient can be related to the free energy of transfer by,

$$\Delta G = -RT \ln K_D \tag{5}$$

where ΔG is the change in Gibbs free energy of solute transfer from the mobile phase to the stationary phase (12), and K_D is the solute distribution coefficient.

The thermodynamic relationship between the solute distribution coefficient and temperature is,

$$\ln K_D = -\Delta H/RT + \Delta S/R \tag{6}$$

where ΔH is the change in ethalpy of solute transfer and ΔS is the change in entropy of solute transfer. The solute distribution coefficient can be related to the solute retention factor by,

$$K_D = C_s/C_m = k'(V_m/V_s) = k'\beta \tag{7}$$

where C_s and C_m are the solute concentrations in the two phases, V_m is the volume of the mobile phase and V_s is the volume of the stationary phase. Their ratio, β, is defined as the phase ratio. Upon substitution of equation 7 into 6 and rearranging yields,

$$\ln k' = -\Delta H/RT + \Delta S/R - \ln \beta \tag{8}$$

The slope of a plot of ln k' against T^{-1} at constant pressure is,

$$[\partial \ln k'/\partial (T^{-1})]_P = -\Delta H/R \tag{9}$$

where ΔH is the enthalpy of solute transfer at constant pressure. Of greater interest is the slope of ln k' versus T^{-1} at constant density,

$$(\partial \ln k'/\partial T^{-1})_\rho = (\partial \ln k'/\partial P)_T (\partial P/\partial T^{-1})_\rho + (\partial \ln k'/\partial T^{-1})_P \tag{10}$$

From Equation 10, the effective enthalpy of solute transfer between the mobile and stationary phases, (ΔH_T) can be obtained at constant density. Substituting Equation 9 into equation 10 yields,

$$(\partial \ln k'/\partial T^{-1})_\rho = \delta + \Delta H = \Delta H_T \tag{11}$$

where δ is a correction term to ΔH containing a thermal expansivity term and $(\partial \ln k'/\partial P)_T (\partial \ln k'/\partial T^{-1})_\rho$. For liquid and gas chromatography, where pressure has a negligible effect on solvent density (HPLC) or is held constant (GLC), δ is zero and Equation 9 is valid.

Therefore the effective enthalpy of solute transfer in SFC can be obtained experimentally for a set of solutes at different densities by determining solute retention as a function of temperature at constant density. The reference state in both the mobile and stationary phase is infinite dilution obeying Henry's Law; therefore, solute-solute interactions do not influence the standard state. The standard state for ΔH_T is a hypothetical one in which the solute is at unit molar concentration having the same properties that it would have at infinite dilution (12).

Solvatochromic Measurement of Supercritical Fluid Solvation Environments. The solvating properties of supercritical fluid mobile phases and how these properties change with pressure (or fluid density) are vital to understanding retention processes in SFC. The measurement of spectroscopic solvent shifts can be used to probe the immediate solvation environment of a solute molecule. This solvatochromic method utilizes a linear solvation energy relationship to correlate solvent effects upon a solute. The Kamlet-Taft solvatochromic scale (one of several which have been developed) relates the solute's susceptibility to the solvent's polarity/polarizability, acid or basic characteristics, and charge transfer effects (13-17). Kamlet and Taft have shown that for UV-visible spectral data, the solvatochromic behavior of the solvent can be correlated by a single parameter, π^*, in the absence of hydrogen bond interactions or charge transfer effects. The π^* parameter correlates solvent effects on the π to π^* electronic transition of the solute molecule, establishing an empirical π^* scale of solvent polarity/polarizability.

The relationship between the solvent's solvatochromic behavior and the susceptibility of the solute in the absence of specific chemical interactions (e.g., hydrogen bonding), is given by,

$$\upsilon_{max} = \upsilon_0 + S\pi^* \tag{12}$$

where υ_{max} is the wave number of maximum absorbance, υ_0 is the reference absorbance maximum of the solute in a standard solvent (cyclohexane) and S is the solute's susceptibility to the solvent's polarity/polarizability (π^*). Using equation 12 the solvent strength

(π^* value) of a supercritical fluid can be related to changes in the wavenumber of maximum absorbance for the solute as a function of pressure, temperature, density and mole fraction of organic co-solvent.

Experimental

The experimental apparatus and methodologies have been described in detail in previous publications. (18-22) The columns used in these studies were fused silica capillaries coated and cross-linked with methylphenylpolysiloxane stationary phases. Detection could be accomplished using either a flame ionization detector or a UV-visible variable wavelength detector, in which a small section of the outer polyimide coating of the capillary had been removed to serve as the sample cell. Chromatographic studies using one component fluids were obtained by condensing the solvent gas into the syringe of a high pressure chromatographic pump. Binary fluid mixtures were prepared by loading predetermined weights of co-solvent in the syringe followed by a known weight of solvent gas from a lecture bottle. The lecture bottle was weighed before and after solvent loading into the syringe to accurately determine the amount of condensed gas added to the co-solvent. The temperature of the columns was controlled with a constant temperature air bath having an accuracy of ±0.1 °C. Solute samples were injected with a Valco C14W HPLC injection valve (0.2 µl rotor volume). The injector was mounted ouside the column oven and was connected to the chosen chromatographic column through a flow splitter. A flow restrictor was connected to the end of the column which controlled the linear velocity of the supercritical fluid mobile phase in the column. The solute retention factor, k', was determined from the ratio of the retention time of the retained solute (t_R) minus the elution time of a non-retained component (t_o) to the elution time of a non-retained component (k' = (t_R-t_o)/t_o).

The probe molecule used for the studies on the solvatochromic behavior of supercritical fluids was 2-nitroanisole, which has an S value of -2.428 ± 0.195 (15) and a reference absorbance maximum of 32,560 cm^{-1} in cyclohexane (v_0). Absorption spectra of 2-nitroanisole in the supercritical fluids were obtained with a Varian Model 2200 spectrophotometer operated in the dual beam mode with an air reference. The gases used as supercritical solvents were SFC grade CO_2, high purity grade N_2O and NH_3, and research grade for the other gases.

The high pressure cell was constructed from stainless steel (SS 304) with dimensions of 8.25 cm wide x 5.0 cm high x 14.0 cm long. The optical path along the axis of the cell was 5.0 cm long x 1.9 cm in diameter with a sapphire window at each end of 2.5 cm in diameter x 1.3 cm thick. The high pressure seal between the stainless steel cell body and the sapphire windows was made using silver-coated, metal C-rings. The volume of the sample cell was ~14.5 cm^3.

Results and Discussion

Prediction of chromatographic retention in SFC using the simple thermodynamic equation outlined in the Theory Section allows one to determine the effect of the bulk macroscopic thermodynamic parameters of partial molar volume of the solute in the mobile and stationary phases upon retention. Equation 3 can be used to calculate solute retention once a k' value for that particular solute at any pressure has been determined experimentally. Prediction of solute retention for naphthalene at 55 °C as a function of pressure with supercritical CO_2 as the fluid mobile phase is shown in Figure 1. The binary interaction parameter used in the calculation of $\bar{v}_1^{mp,\infty}$ from the Peng-Robinson EOS was $k_{12} = 0.040$, which was regressed from naphthalene solubility data in CO_2 at 55 °C (23). As shown in Figure 1, there is an excellent correlation between the experimental retention data and the theoretical prediction of retention for naphthalene at 55 °C over a wide range of pressure. The value of $\bar{v}_1^{sp,\infty}$ used for the calculation of naphthalene retention was -125 cm^3/mole; although the relative magnitude of this term may be questionable, the value suggests an attractive interaction between the solute and the bonded polymeric stationary phase. The pressure independence of $\bar{v}_1^{sp,\infty}$ may or may not be valid at higher fluid densities and for stationary phases that are solvated by the supercritical fluid. For low densities and inert stationary phases the thermodynamic model's predictions are quite accurate. Figure 2 shows the same experimental results as used in Figure 1, with solute retention for naphthalene plotted as a function of density for CO_2 at 55 °C. Solute retention was calculated using equation 4 and the Peng-Robinson EOS. Once again, excellent correlation is seen between the experimental retention and the theoretical prediction using the same binary interaction parameters and $\bar{v}_1^{sp,\infty}$.

For non-polar solutes, the simple thermodynamic model described can be used to predict solute retention as a function of pressure or density at constant temperature. This model will be valid for those solute-solvent systems where specific interactions (e.g., hydrogen bonding or charge transfer complex formation) are negligible or non-existent. In this case, the Peng-Robinson EOS provides an adequate description of the solute-solvent interactions, which over the limited pressure range studied, dominates the solute retention process. The ability to calculate solute retention in SFC would lead to increased speed and efficiency in selecting appropriate experimental conditions for a particular chromatographic separation, maximizing resolution as a function of separation time. The study of solute-stationary phase interactions through this model remains an intriguing possibility to explore.

The thermodynamics of solute retention in SFC can be studied either as a function of pressure or temperature. The bulk macroscopic thermodynamic parameters affecting solute retention as a function of pressure have already been discussed and shown to be the partial molar volume of the solute in the mobile and stationary phases ($\bar{v}_1^{mp,\infty}$ and $\bar{v}_1^{sp,\infty}$). At constant pressure, more specifically constant density, the bulk macroscopic thermodynamic variable which is temperature dependent, is the effective enthalpy of solute transfer between the mobile and stationary phase (ΔH_T). Chester and Innis (24) have studied solute retention as a function of temperature at constant pressure and arrived at a thermodynamic explanation of observed retention in SFC relating solute volatility and fluid-solute solvation contributions to retention. The study of ΔH_T at constant density as a function of temperature provides a more direct route to physical insights into the retention mechanism of SFC.

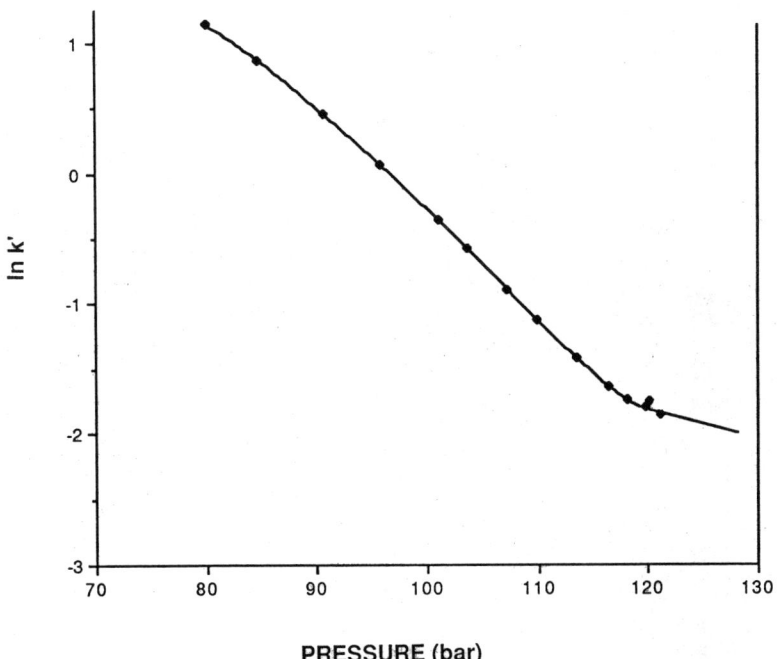

Figure 1. Plot of ln k' versus pressure (bar) for naphthalene retention at 55 °C in CO_2 on SE-54, $\bar{v}_1^{sp,\infty}$ = -125 cm^3/mole, (◆) experimental data (—) calculated retention, binary interaction parameter = 0.040.

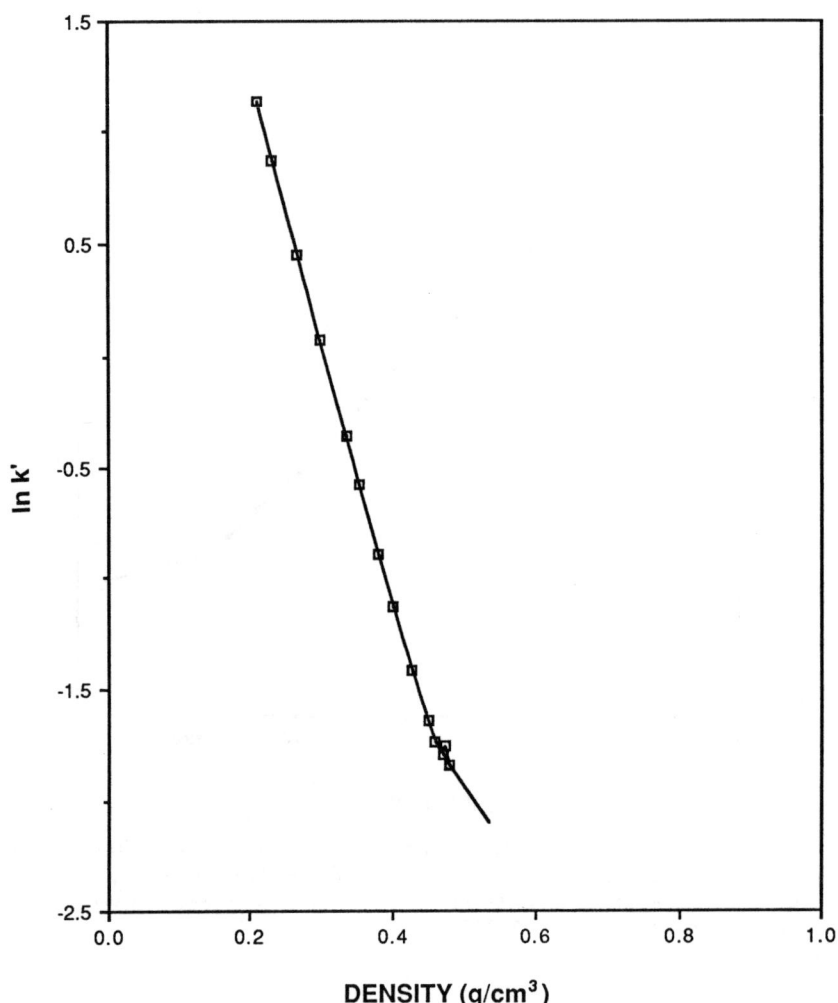

Figure 2. Plot of ln k' versus density (g/cm^3) for naphthalene retention at 55 °C in CO_2 on SE-54, $\bar{v}_1{}^{sp,\infty}$ = -125 cm^3/mole, (■) experimental data (—) calculated retention, binary interaction parameter = 0.040.

For pure fluid solvents Lauer et al. (25) have reported average values for ΔH_T of -6 kcal/mole for selected solutes with supercritical CO_2 and N_2O at a density of 0.80 gm/cm^3 with PRP-1 (a styrene-divinylbenzene copolymer) micro-particulate packed column. Yonker and Smith (18) have reported effective enthalpy values of -5.9 to -8.4 kcal/mole and -3.9 to -8.6 kcal/mole for select solutes on OV-17 and SE-54 capillary columns, respectively, for CO_2 densities between 0.20 and 0.50 g/cm^3. Figure 3 is a plot of ΔH_T as a function of density for two different stationary phases, SE-54 and OV-17. The ΔH_T values for the n-alkane, heptadecane, on SE-54 with CO_2 show a nearly linear dependence upon fluid density, while the ΔH_T values for OV-17 with CO_2 asymptotically approach a limiting value at higher densities. This behavior for OV-17 may be attributed to possible solvation or swelling of the bonded polymeric phase by CO_2, which would affect ΔH_T. The greatest solvation occurs at higher densities, resulting in a less negative ΔH_T value (partition-like mechanism). The phenomenon of bonded phase swelling or solvation has been reported for pure supercritical fluids by Sie et al. (26) and Springston et al. (27).

Modified (mixed) fluid systems are of interest because many of the attributes of a more polar fluid can be obtained, while retaining a mild critical temperature for the mixture. The retention characteristics of binary supercritical fluid solutions containing CO_2 and an organic co-solvent as a function of temperature at constant density have been studied. In addition, various binary fluid densities and co-solvent mole fractions were studied. The densities of the binary fluids were determined as outlined by Yonker and Smith in their work on composition gradients in SFC (28). Figure 4 shows the dependence of ΔH_T upon modifier concentration for the binary fluid CO_2/2-propanol at two densities, 0.44 g/cm^3 and 0.39 g/cm^3, using the polar solute myristophenone with an SE-54 bonded polymeric stationary phase. Two points are apparent in Figure 4: (1) a minimum is observed in ΔH_T between a co-solvent mole fraction of 0.0 and 0.05, and (2) ΔH_T becomes more negative as co-solvent mole fraction increases. The thermodynamic behavior (ΔH_T) of the binary fluid solvent (Figure 4) contrasts with that seen for the pure fluid (Figure 3). Once again this is most likely due to the dynamic process of bonded stationary phase solvation by the added co-solvent modifier. Small amounts of organic co-solvents have been shown to have a large effect on solute retention in SFC (20, 28-33) (Yonker, C. R.; McMinn, D. G.; Wright, B. W.; Smith, R. D. J. Chromatogr, in press). As the stationary phase becomes solvated by 2-propanol, the SFC retention mechanism assumes a greater partition-like characteristic compared to the ΔH values reported in liquid chromatography (34,35). As the mole fraction of organic co-solvent increases, ΔH_T becomes more negative. The thermodynamics of solute retention at constant density for SFC are complex due to the dynamics of processes relevant to solute transfer between the mobile and stationary phases. The retention mechanism in SFC suggested by the studies at constant density is consistent with the observation that ΔH_T becomes more negative as the density decreases, ultimately approaching the enthalpy values reported for gas chromatography (36-38). Therefore, it is suggested that for the system studied the retention mechanism varies as a function of density between an adsorption-like process as seen in gas chromatography to a partition-

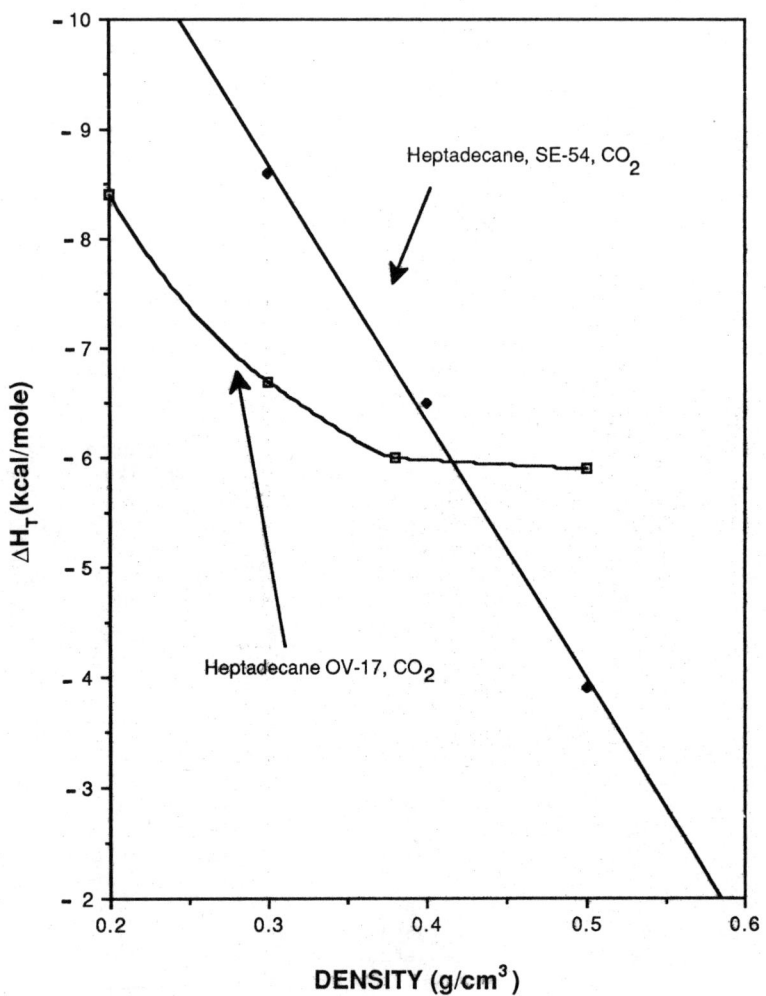

Figure 3. Plot of ΔH_T versus density (g/cm^3) for heptadecane on OV-17 (☐) and SE-54 (◆).

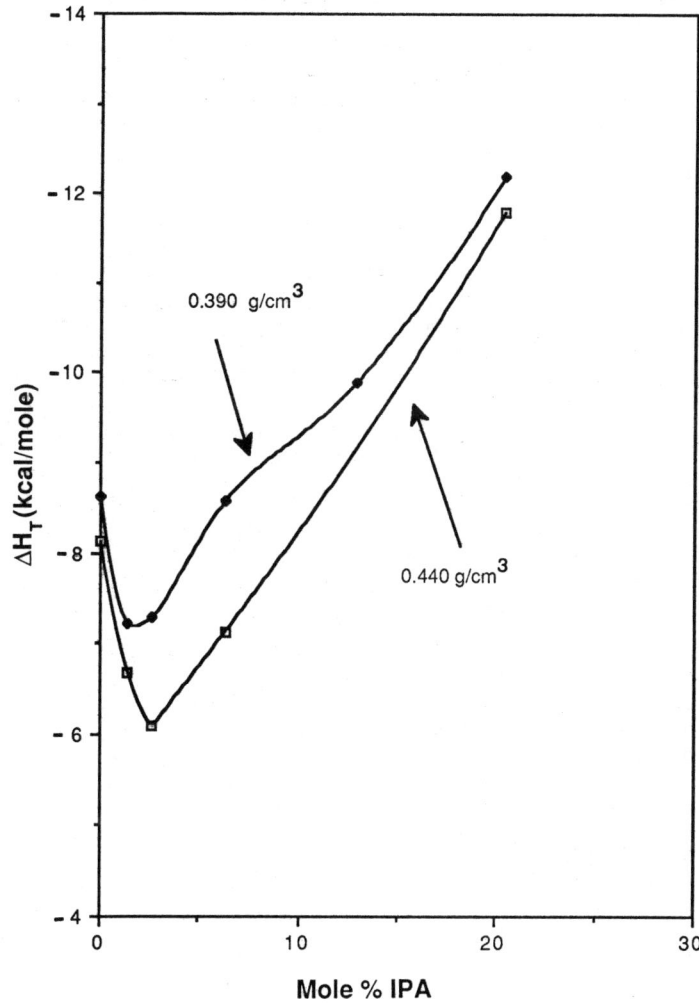

Figure 4. Plot of ΔH_T versus mole % 2-propanol (mole % IPA) for myristophenone on SE-54 at the densities of (□) 0.440 g/cm^3 and (◆) 0.390 g/cm^3.

like process as fluid density and solvation of the bonded polymeric stationary phase increases (18,24). For binary fluid mixtures, ΔH_T displays a minimum as a function of modifier concentration due to the solvation of the bonded stationary phase by the organic co-solvent. Clearly, the enrichment (or modification) of the stationary phase by supercritical fluid co-solvents has a prominent effect on the retention mechanism in SFC. More research is being undertaken to explore the solvation of various polymeric phases by organic co-solvents in an effort to define its effect on the retention mechanism in SFC.

The solute-solvent interactions independent of the stationary phase are obviously of major importance to understanding the retention mechanism in SFC. The solvatochromic behavior of supercritical fluid probes has been used to investigate the specific and non-specific solute-solvent interactions occurring in a fluid as a function of density at different temperatures. Yonker et al. (19, 39,40) have studied various supercritical fluids over a wide range of density and temperature conditions with the aim of allowing intercomparison of supercritical fluid solvents as well as comparison with conventional liquid solvents. Kim and Johnston (41) have also reported solvatochromic data using pure and binary mixtures of supercritical fluids.

The π^* polarizability/polarity values determined for various supercritical fluids of general interest to SFC are shown in Figure 5. From this figure one can draw some interesting conclusions: 1) ammonia is the most polar supercritical fluid studied, whose solvent strength (ignoring specific chemical interactions) varies from that of hexane to ethanol as a function of density, 2) CO_2 and N_2O show similar solvent strengths as a function of density over the entire regime studied [this is not surprising when considering a corresponding states argument for these fluids since their critical conditions are so similar (42)], and 3) SF_6 and ethane are non-polar fluids and display low polarity on the π^* scale due to their polarizability dominating in solute-solvent interactions. It should be noted that these general observations are consistent with our experience in utilizing these fluids for SFC.

The solvatochromic technique directly probes the solute-solvent interactions in the cybotactic region of the solute. The cybotactic region is defined as the cage of solvent molecules whose structure about the solute molecule is determined by the presence of the solute. The immediate solvation sphere about the solute molecule for binary fluid solvents may be anticipated to show preferential enrichment of the organic co-solvent due to the greater attractive forces. Figure 6 contains the spectroscopic data with values based on equation 12 for a CO_2/2-propanol (IPA) binary fluid solvent using 2-nitroanisole as the solvatochromic probe molecule at various concentrations of IPA at 44 °C. It should be noted that the addition of a polar modifier such as IPA to CO_2 probably results in specific interactions (hydrogen bonding) with the solvatochromic probe. Thus the π^* values calculated from equation 12 do not strictly conform to the approach of Kamlet and Taft, which would require a more rigorous approach invoking additional solvatochromic parameters. However, the present approach provides a useful description of the fluid solvating power and a basis for more quantitative determinations. The results in Figure 6 show a shift towards more polar π^* values on addition of

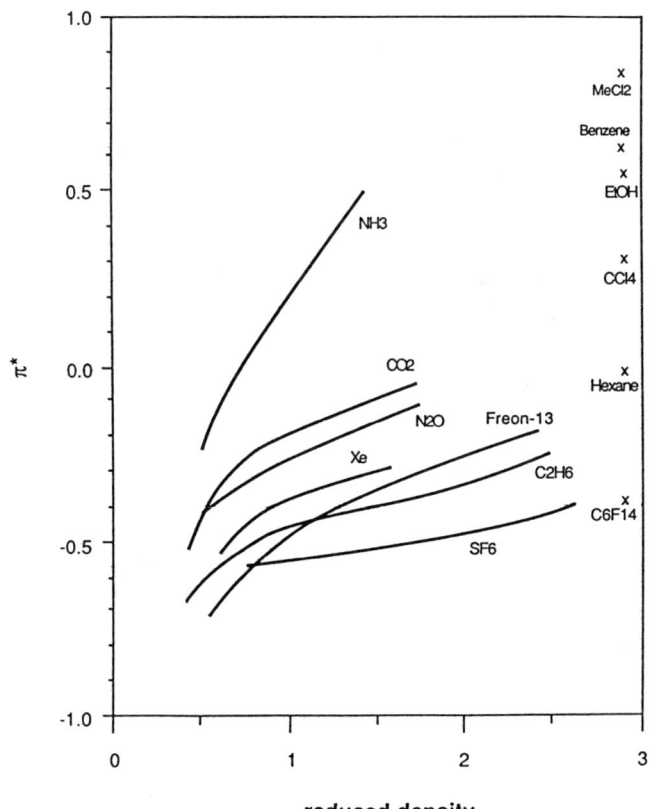

Figure 5. Plot of π^* versus reduced density for NH_3, CO_2, N_2O, Xe, Freon-13, C_2H_6 and SF_6 with 2-nitroanisole at a reduced temperature of 1.03. The π^* values for liquid methylene chloride ($MeCl_2$), benzene, ethanol (EtOH), carbon tetrachloride (CCl_4), hexane and per-fluorohexane (C_6F_{14}) are included for comparison.

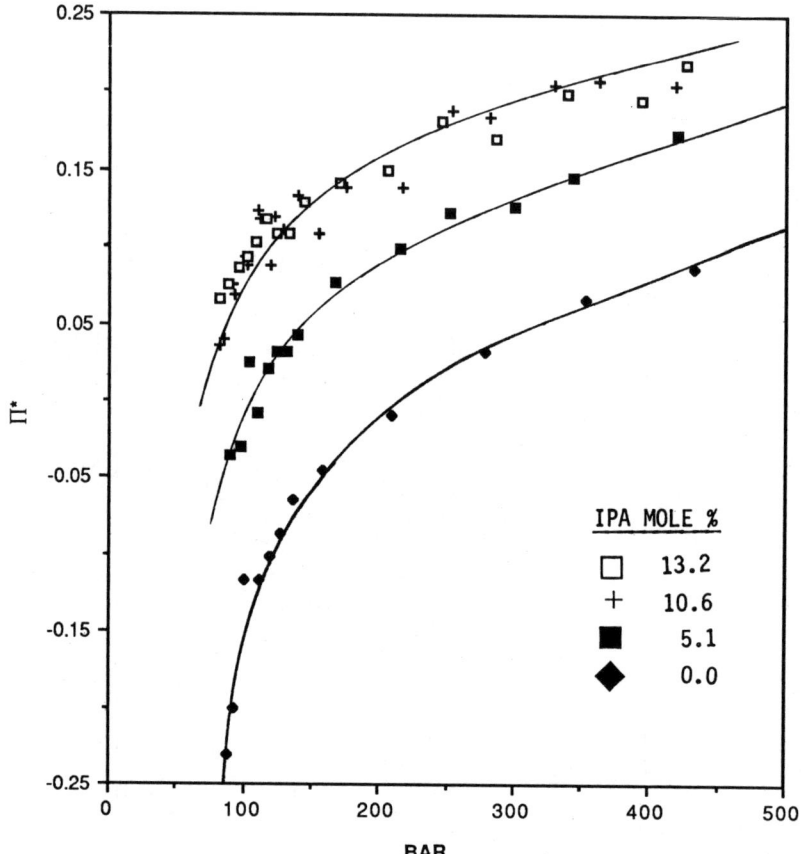

Figure 6. Plot of π^* versus pressure (bar) for various mole % 2-propanol (IPA) in CO_2 with 2-nitroanisole at 44 °C.

a small concentration of IPA. The π* value becomes more positive, approaching the limit of the value for pure IPA. An enrichment of IPA, beyond the 5.1 mole %, is seen in the cybotactic region about 2-nitroanisole (Yonker, C. R.; Smith, R. D. *J. Phys. Chem.* in press), which contributes to the solvatochromic shift. At 44 °C, the 10.6 and 13.2 mole % CO_2/IPA are subcritical liquids at the pressures studied in Figure 6 (43). Therefore, the solvatochromic shift reflects the saturation of the cybotactic region of the solute by the subcritical liquid co-solvent and no significant difference is seen between the extent of the spectral shifts for 10.6 and 13.2 mole % IPA.

Supercritical fluid solvent strength will affect solute-solvent interactions in the fluid mobile phase which will impact on solute retention in SFC. As the concentration of polar organic co-solvent increases the solvent strength (polarity) also increases (see Figure 6) and retention of a polar solute would be expected to decrease. Figure 7 is a plot of solute retention (ln k') as a function of density for CO_2/IPA for various concentrations of the organic co-solvent. The solute studied was 9-phenanthrol, which could interact with the polar IPA. At constant density, solute retention decreases as a function of IPA concentration. These experiments were performed at 127 °C with SE-54 as the stationary phase. The results shown in Figure 7 are entirely consistent with the increased solvent strength (increased polarity) demonstrated by the solvatochromic studies for binary supercritical fluid solvents.

The solvatochromic method can be used to probe the solute's cybotactic region, to determine the effect of solvation on the physicochemical properties of the solute. This information will lead to a more fundamental understanding of the role of the fluid mobile phase in the retention mechanism for SFC. Binary fluids can show an enrichment in the solute cybotactic region, which can alter the selectivity of the separation process, through changes in solute-solvent interactions as a function of density, temperature or co-solvent concentration. Important questions which remain to be addressed include the effect of solvent modifier saturation of the cybotactic region (at high concentrations) upon retention. The physical properties of mixed fluids (i.e., diffusion coefficients) relevant to SFC remains essentially unexplored, although the local concentration of the modifier in the cybotactic region might result in substantially lower diffusion rates. Finally, as indicated earlier, the effect of modifiers on SFC stationary phases is a complex process and requires significant further study.

Conclusion

Retention in SFC is a complex function of temperature, pressure, density and solvent modifier concentration and a more complete understanding of these phenomena should directly benefit the development of SFC. The dynamic processes of stationary phase solvation and solute-solvent interactions in the stationary and mobile phases, respectively, impact on solute retention in SFC. The study of the retention process necessitates a multi-dimensional approach to understand the basic physicochemical processes underlying solute retention in SFC. The discussion in this chapter outlines three interrelated areas of study, probing specific areas of solute

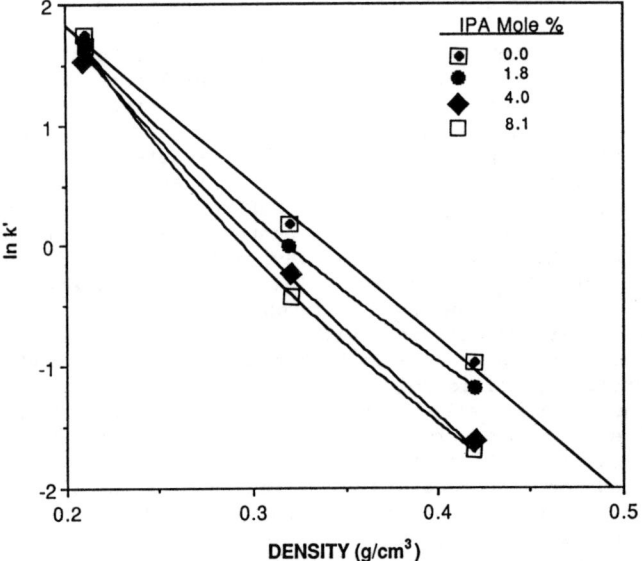

Figure 7. Plot of ln k' versus density (g/cm^3) for various mole % of 2-propanol (IPA) in CO_2 with 9-phenanthrol on SE-54 at 127 °C.

retention. Prediction of solute retention using a simple thermodynamic model allows one to study the effect of the bulk macroscopic thermodynamic parameters of $\bar{v}_1^{mp,\infty}$ and $\bar{v}_1^{sp,\infty}$ and their role in retention. The attractive interaction between the solute and the bonded stationary phase is an interesting area of future research which will impact on the basic understanding of the retention mechanism for SFC. The thermodynamic studies of retention as a function of temperature at constant density presents the capability of studying SFC retention from gas-like densities to liquid-like densities, bridging the gap between gas and liquid chromatography. Solvation of the stationary phase has a possible impact on ΔH_T, the importance of this to retention in SFC is still under active study. The solvatochromic technique presents the ability to study the fluid mobile phase solvent-solute interactions independent of the role of the stationary phase in the retention process for SFC. The direct study of the cybotactic region of the solute in supercritical fluids allows one to bridge the gap between gas phase and condense phase regions by studying cluster formation and solvation of the solute by the supercritical fluid. The combination of these three techniques and the fundamental physicochemical information obtained thus far describe more coherently the retention mechanism for SFC.

Acknowledgment

This work has been supported by the U. S. Department of Energy, Office of Basic Energy Sciences, under Contract DE-AC06-76RLO 1830.

Literature Cited

1) Schmitt, W. J.; Reid, R. C. _J. Chem. Eng. Data_ 1986, _31_, 204.
2) Dobbs, J. M.; Wong, J. M.; Johnston, K. P. _J. Chem. Eng. Data_ 1986, _31_, 303.
3) Dobbs, J. M.; Wong, J. M.; Lahiere, R. J.; Johnston, K. P. _Ind. Eng. Chem. Res._ 1987, _26_, 56.
4) McHugh, M. A.; Krukonis, V. J. _Supercritical Fluid Extractions: Principles and Practice_; Butterworths: Boston, 1986; Chapter 11.
5) Hamann, S. P. _High Pressure Physics and Chemistry_; Bradley, R. S., Ed.,; Academic Press: New York, 1963; Vol. 2, Chapter 8.
6) Van Wasen, U.; Schneider, G. M. _Chromatographia_ 1975, _8_, 274.
7) Yonker, C. R.; Gale, R. W.; Smith, R. D. _J. Phys. Chem._ 1987, 91, 3333.
8) McHugh, M. A.; Paulaitis, M. E. _J. Chem. Eng. Data_ 1980, _25_, 326.
9) Wu, P. C.; Ehrlich, P. _AIChE J._ 1973, _19_, 533.
10) Ziger, D. H. PhD Thesis, University of Illinois, Urbana, 1984.
11) Peng, D. Y.; Robinson, D. B. _Ind. Eng. Chem. Fundam._ 1976, _15_, 59.
12) Meyer, E. F. _J. Chem. Educ._ 1973, _50_, 191.
13) Kamlet, M. J.; Abboud, J-L. M.; Taft, R. W. _Prog. Phys. Org. Chem._ 1980, _13_, 485.
14) Kamlet, M. J.; Abboud, J-L. M.; Abraham, M. H.; Taft, R. W. _J. Org. Chem._ 1983, _48_, 2877.
15) Kamlet, M. J.; Doherty, R. M.; Taft, R. W.; Abraham, M. H. _J. Am. Chem. Soc._ 1983, _105_, 6741.

16) Kamlet, M. J.; Abboud, J-L. M.; Taft, R. W. *J. Am. Chem. Soc.* 1977, **99**, 6027.
17) Abboud, J-L. M.; Kamlet, M. J.; Taft, R. W. *J. Am. Chem. Soc.* 1977, **99**, 8325.
18) Yonker, C. R; Smith, R. D. *J. Chromatogr.* 1986, **351**, 211.
19) Yonker, C. R.; Frye, S. L.; Kalkwarf, D. R.; Smith, R. D. *J. Phys. Chem.* 1986, **90**, 3022.
20) Yonker, C. R.; Smith, R. D. *J. Chromatogr.* 1986, **361**, 25.
21) Wright, B. W.; Smith, R. D. *Chromatographia*, 1984, **18**, 542.
22) Smith, R. D.; Kalinoski, H. T.; Udseth, H. R.; Wright, B. W. *Anal. Chem.* 1984, **56**, 2476.
23) Mackay, M. E.; Paulaitis, M. E. *Ind. Eng. Chem. Fundam.* 1979, **18**, 149.
24) Chester, T. L.; Innis, D. P. *J. High Resol. Chromatogr. & Chromatogr. Commun.* 1985, **8**, 561.
25) Lauer, H. H.; McManigill, D.; Board, R. D. *Anal. Chem.* 1983, **55**, 1370.
26) Sie, S. T.; Van Beersum, W.; Rijnders, G.W.A. *Sep. Sci.* 1966, **1**, 459.
27) Springston, S. R.; David P.; Steger, J.; Novotny, M. *Anal. Chem.* 1986, **58**, 997.
28) Yonker, C. R.; Smith, R. D. *Anal. Chem.* 1987, **59**, 727.
29) Crowther, J. B.; Henion, J. D. *Anal. Chem.* 1985, **57**, 2711.
30) Blilie, A. L.; Greibrokk, T. *Anal. Chem.* 1985, **57**, 2239.
31) Schmitz, F. P.; Klesper, E. *Polym. Commun.* 1983, **24**, 142.
32) Schmitz, F. P.; Hilgers, H.; Klesper, E. J. *J. Chromatogr.* 1983, **267**, 267.
33) Blilie, A. L.; Greibrokk, T. *J. Chromatogr.* 1985, **349**, 317.
34) Grushka, E.; Colin, H.; Guiochon, G. *J. Chromatogr.* 1982, **248**, 325.
35) Knox, J. H.; Vasvori, G. *J. Chromatogr.* 1973, **83**, 181.
36) Meyer, E. F.; Stec, K. S.; Hotz, R. D. *J. Phys. Chem.* 1970, **77**, 2140.
37) Martire, D. E.; Tewari, Y. B.; Sheridan, J. P. *J. Phys. Chem.* 1970, **74**, 2345.
38) Martire, D. E.; Tewari, Y. B.; Sheridan, J. P. *J. Phys. Chem.* 1970, **74**, 2363.
39) Frye, S. L.; Yonker, C. R.; Kalkwarf, D. R.; Smith, R. D. In *Supercritical Fluids, Chemical and Engineering Principles and Applications*; Squires, T. G.; Paulaitis, M. E., Eds.; ACS Symposium Series, No. 329; American Chemical Society: Washington, DC, 1987; Chapter 3.
40) Smith, R. D.; Frye, S. L.; Yonker, C. R.; Gale, R. W. *J. Phys. Chem.* 1987, **91**, 3059.
41) Kims, S.; Johnston, K. P. In *Supercritical Fluids, Chemical and Engineering Principles and Applications*; Squires, T. G.; Paulaitis, M. E., Eds.; ACS Symposium Series, No. 329; American Chemical Sociey: Washington, DC 1987; Chapter 4.
42) Schoenmakers, P. J. *J. Chromatogr.* 1984, **315**, 1.
43) Radosz, M. *J. Chem. Eng. Data* 1986, **31**, 43.

RECEIVED October 9, 1987

Chapter 10

Capillary Supercritical Fluid Chromatography

Use for the Analysis of Food Components and Contaminants

B. E. Richter, M. R. Andersen, D. E. Knowles, E. R. Campbell, N. L. Porter, L. Nixon, and D. W. Later

Lee Scientific, Inc., 4426 South Century Drive, Salt Lake City, UT 84123

> Capillary supercritical fluid chromatography has been demonstrated as a viable alternative for the analysis of food components which are sensitive to temperature such as flavors and fragrances. Supercritical fluids have long been recognized for their unique solvating characteristics. One of the most common uses of supercritical fluids is for the extraction of components of interest from natural materials (i.e., caffeine from coffee or oil from soybeans). Early in its development, supercritical fluid chromatography (SFC) was used for the analysis of natural materials such as flavors and other food components because the technique is well suited for the analysis of compounds which thermally degrade. In this paper, the use of capillary SFC for the analysis of food components is discussed. Examples of the capillary SFC analysis of fats and flavors as well as food contaminants such as pesticides are presented.

There are many literature reports on industrial uses of supercritical fluids, and many patents have been issued for various uses of supercritical fluids as solvents in extraction processes. Randall (1) prepared an excellent review on the uses and patents issued up to 1982 in the area of supercritical fluid extraction and chromatography. Many of these applications deal with natural products such as flavors, fragrances, oils and fats, or the removal of unwanted components from natural materials such as caffeine from coffee or tea and nicotine from tobacco.

The first use of supercritical fluids as mobile phases in chromatography was reported in 1962 by Klesper et al. (2). The interest in the use of supercritical fluids grew slowly at first, probably due in large measure to the concurrent rapid development of high performance liquid chromatography (HPLC) as well as some technical difficulties which impeded the introduction of commercial SFC instrumentation. Interest began to grow rapidly in

0097-6156/88/0366-0179$06.00/0
© 1988 American Chemical Society

the early 1980's as the first commercial packed column SFC instrument was introduced. Shortly following this event, the first report appeared in the literature on capillary column SFC by Novotny and Lee (3). Since its introduction, SFC technology has grown rapidly in use and application. Subsequently, the first commercial capillary column SFC instrumentation was introduced in 1986.

The use of capillary SFC for the analysis of food components was first reported in 1984 by Chester (4) and then in 1985 by White (5). Both groups showed the analysis of mono-, di-, and triglycerides at low operating temperatures using carbon dioxide as the mobile phase to avoid degradation of the unsaturated fatty acid groups. Additional work by Chester and co-workers also has shown the analysis of sucrose polyesters (6), oligo- and polysaccharides (7), and polyglycerol esters (8). Capillary SFC using carbon dioxide as the mobile phase can be operated near room temperature and is thus well suited for the analysis of food components which can thermally degrade, such as citrus oils (9).

In this paper, work is presented showing capillary SFC analyses of a soybean oil, a hops extract, a celery seed oleoresin, and an essential citrus oil. In addition, results showing the determination of pesticide residues in a parsley sample by capillary SFC are presented.

Experimental

Supercritical fluid grade carbon dioxide (Scott Specialty Gases, Plumsteadville, PA) was used as the carrier fluid. A Lee Scientific Model 501 supercritical fluid chromatograph equipped with a flame ionization detector (FID) and a nitrogen-phosphorus detector (NPD) was the instrument utilized for these studies. Fused silica capillary columns (50 µm i.d.) were employed for all the experiments. Three column types with stationary phases of three different polarities were used: SB-Methyl-100, SB-Biphenyl-30 and Carbowax 20M (0.25 µm films). Frit restrictors were used to maintain pressure and proper flow rates in the column. The restrictor was connected to the end of the column via a zero dead-volume union. The end of the restrictor was positioned in the detector at 1 mm below the end of the flame jet. The detector was operated at 325-350°C with nitrogen make-up gas at 25 mL/min. Split injection was used in these experiments with 0.2 µL injection rotor and a split ratio of approximately 10:1.

The citrus oil samples were analyzed neat. Other samples were prepared by dissolving in an appropriate solvent such as methylene chloride, methanol or acetone (HPLC grade).

Results and Discussion

Capillary SFC can be used to analyze triglycerides which contain unsaturated fatty acid groups that are susceptible to degradation. Figure 1 shows the analysis of a soybean oil sample on a 10 m SB-Methyl-100 column using CO_2 as the mobile phase at 200°C with an FID. Identification of the compounds was achieved by comparing the retention times of sample components with standard compounds.

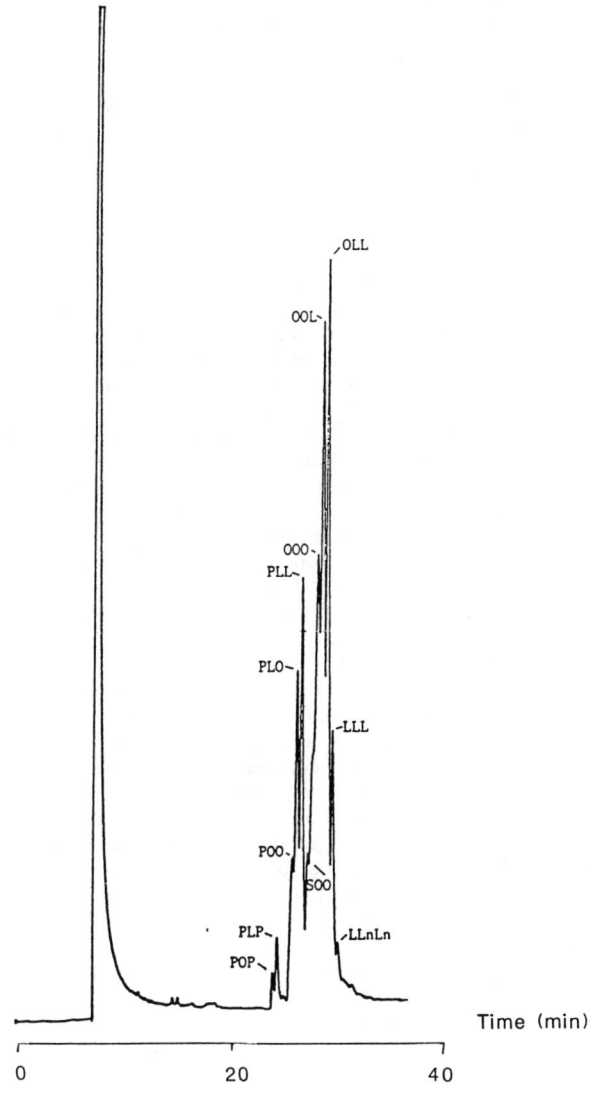

Figure 1. SFC chromatogram of soybean oil sample. Carbon dioxide mobile phase at 200°C, 10 m x 50 μm i.d. SB-Methyl-100 column, linear density programmed at 0.005 g/mL/min, FID at 375°C. Fatty acid group identification: P-Palmitic, O-Oleic, S-Stearic, L-Linoleic, Ln-Linolenic.

The methyl phase column does not give adequate separation of the various triglycerides (Figure 1). However, if the same analysis is done using the more polar Carbowax 20M stationary phase, then better resolution of the components is achieved as shown in Figure 2. It should be noted that these SFC chromatograms were obtained at fairly mild conditions, 200°C, well within the conditions recommended by Proot et al. (150-200°C) for the analysis of triglycerides by capillary SFC (10). Triglycerides which contain unsaturated fatty acid groups can degrade at the elevated temperatures necessary for their elution under GC conditions (350-400°C). Thermal degradation ranging from 3 to 15% has been reported during the analysis of triglycerides by capillary GC (11). Mares et al. showed that quantitative analysis of trace levels of triglycerides can be difficult because of the "low recovery of these compounds" (12). The analysis of triglycerides by HPLC has been thoroughly investigated (13), but it suffers some drawbacks. The absence of a strong chromophore means that UV absorption detection of triglycerides yields very poor sensitivity. Refractive index detection is more frequently used for triglyceride analysis, but this detector cannot be used in conjunction with gradient elution which is necessary for complete resolution of complex mixtures of triglycerides. HPLC separation of triglycerides with laser light scattering detection has been investigated and is gaining popularity (14). Capillary SFC does not suffer from the problems of detector sensitivity or compatibility as does HPLC or the problems of degradation encountered in GC. Triglyceride analysis by capillary SFC does show great promise.

Figure 3 shows the capillary SFC analysis of a liquid CO_2 extract of hops. Compounds eluting in Region A are classified as _-acids and iso- -acids. Examples of these compounds are cohumulone, humulone, and isocohumulone, among others. Positive identification of each of the peaks in the chromatogram was difficult because of the unavailability of pure standards. However, standards of the highest available purity were analyzed under identical conditions to aid in component identification. The information gained from these analyses was used to identify the respective component groupings in the chromatogram. Compounds eluting in Region B are classified as -acids; examples of these compounds include colupulone and adlupulone. All of these compounds are very sensitive to temperature which makes analysis by GC difficult; HPLC does not have sufficient chromatographic efficiency to adequately resolve the compounds as well as can be done by SFC (15,16). Clearly, capillary SFC can easily be used as a fingerprint method for quality control of the flavor components in hops and hops extracts.

Celery seed oleoresin presents a unique analytical challenge. The flavor and fragrance components are easily degraded with temperature (17). However, higher temperatures are required to perform a complete analysis of the sample by GC because triglycerides are present. These triglycerides can also degrade either in hot injection ports or in the columns themselves if oxygen or catalytic impurities are present. Figure 4 shows two chromatograms obtained from the analysis of a celery seed oil.

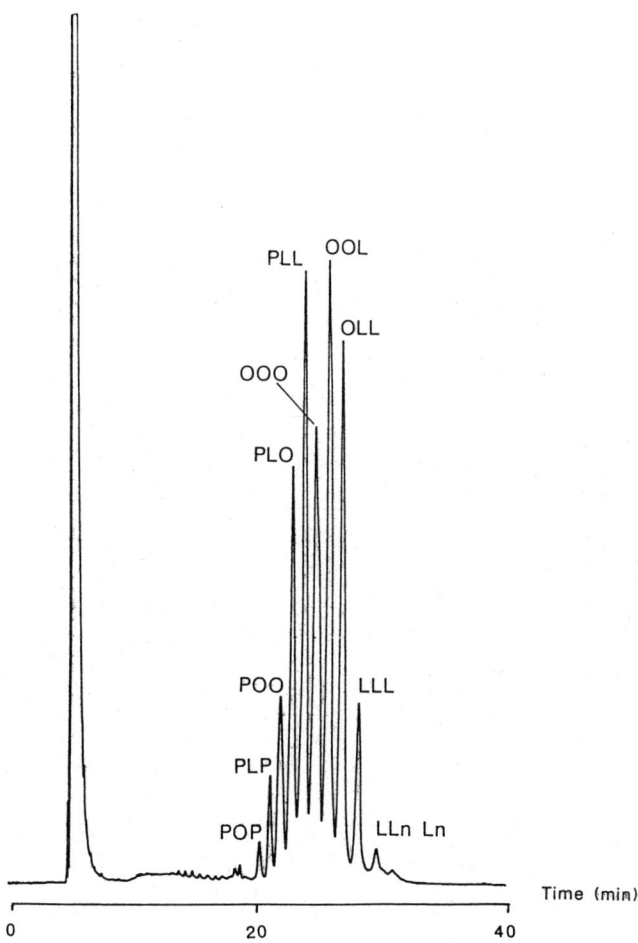

Figure 2. SFC chromatogram of soybean oil sample. CO_2 at 200°C, 10 m x 50 μm i.d. Carbowax 20M column, linear density programmed, at 0.005 g/mL/min, FID at 375°C.

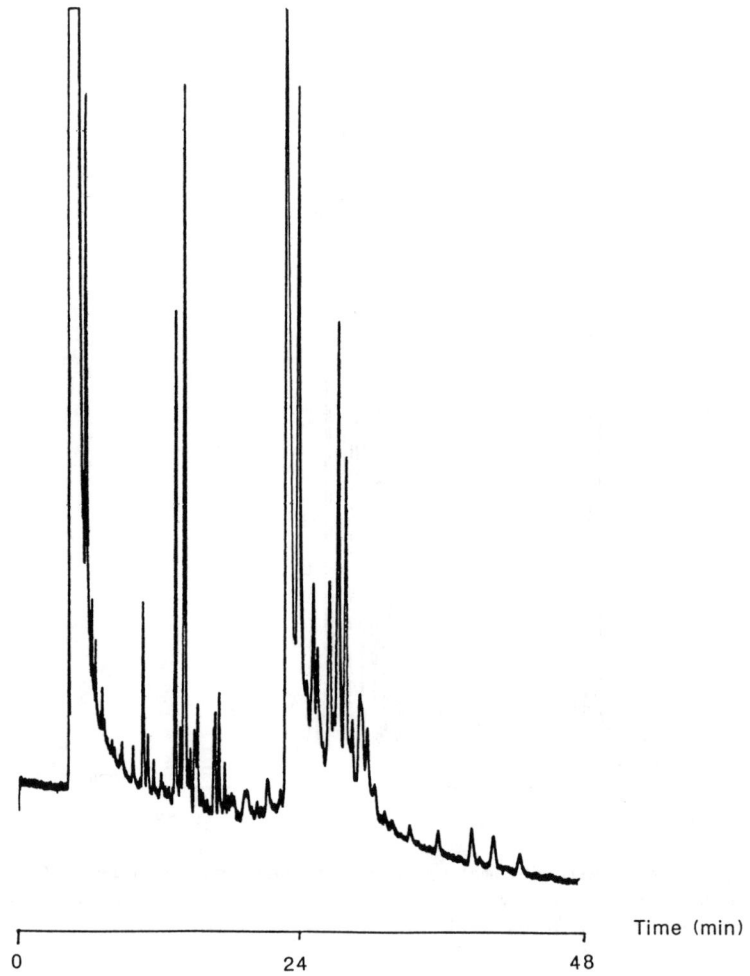

Figure 3. SFC chromatogram of liquid CO_2 extract of hops. CO_2 mobile phase at 100°C, 10 m x 50 μm i.d. SB-Biphenyl-30 column, linear density programmed, FID at 350°C.

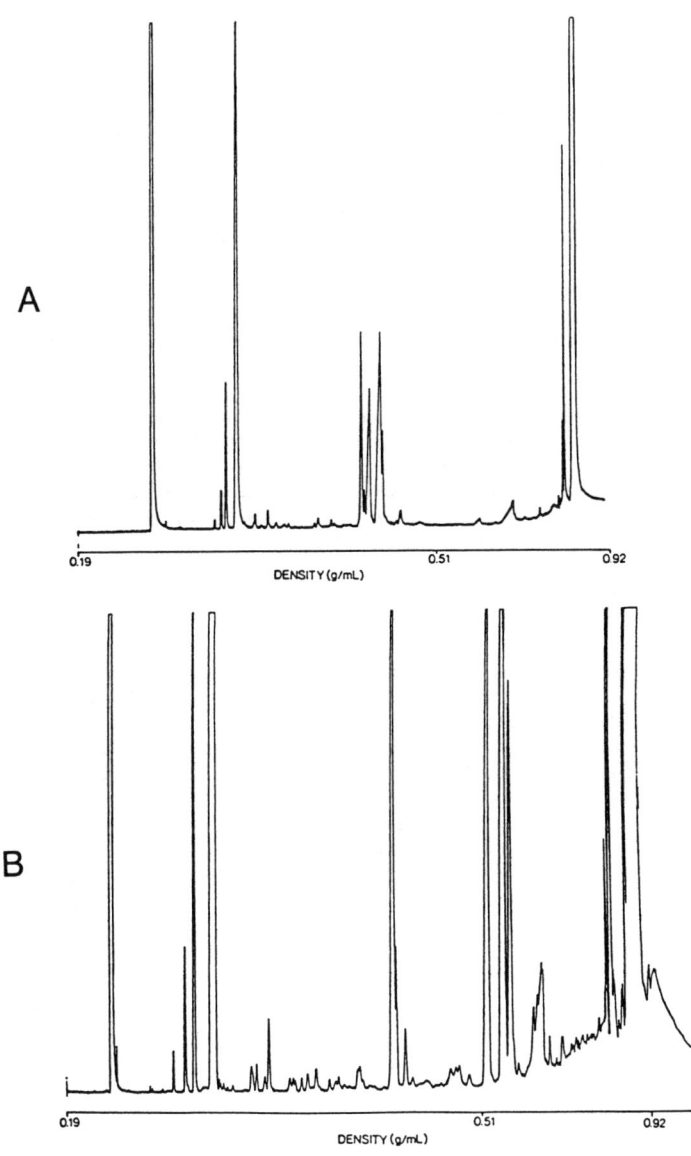

Figure 4. SFC chromatograms of celery seed oil. CO_2 at 50°C, FID at 325°C, multi-ramp density programmed. Top chromatogram (A)--4 m x 50 μm i.d. SB-Methyl-100 column; bottom (B)--4 m x 50 μm i.d. SB-Biphenyl-30 column.

The top chromatogram was obtained on a methyl column, and the bottom chromatogram was obtained on a biphenyl column. Clearly, the biphenyl stationary phase exhibits very different selectivity from the methyl. As points of reference, the peaks eluting in the middle of the chromatograms are alkylated phthalides, and the peaks eluting at the end of the chromatogram are triglycerides. Other components present include terpenes and sesquiterpenes. All these classes of compounds which have widely varying chemical functionalities and molecular weights are separated in a single chromatographic analysis. It should be mentioned that this data was obtained at an operating temperature of 50°C. The probability of thermal degradation at this temperature is greatly reduced. These low temperature analyses can be performed by SFC with the added advantage of using the FID which offers universal and sensitive detection, an option not available in HPLC.

Even though citrus oils are traditionally analyzed by GC, there has been concern raised about this practice because some of the components in these oils can thermally degrade (9). Figure 5 shows two chromatograms obtained from the analysis of a cold pressed grapefruit oil. The sample was analyzed isothermally at 50 and 100°C on a biphenyl stationary phase column. At 50°C, the lighter molecular weight materials which elute early in the chromatogram are well separated while the later eluting components are not well resolved. At 100°C, the later eluting peaks are separated better, and the analysis time is shorter. However, the early eluting peaks are not well resolved but are bunched together and look similar to a solvent peak that would be obtained by GC analysis. If a higher temperature is used (150°C), then the analysis time is again shortened and a slight improvement in the separation of the later eluting peaks is achieved (Figure 6A). As was expected, resolution in the first part of the chromatogram is extremely poor. The bottom chromatogram (Figure 6B) demonstrates that simultaneously combining density programming with temperature programming in a single run achieves a unique separation capability. The initial temperature of this analysis was held at 50°C and the density was at 0.19 g/mL. This allowed the separation of the early eluting peaks. Then as the density was programmed, the temperature was simultaneously programmed to 150°C. This combination of temperature and density programming resulted in the shortest analysis time with the best resolution of all the components, from the early to the later eluting compounds. Even the upper temperature of 150°C is much lower than the temperatures normally used to analyze these samples by GC. Capillary SFC can serve as an alternative to the analysis of flavors and fragrances by GC if the thermal stability of the components of interest is in question.

The analysis of pesticide residues and metabolites in foods can be a difficult problem because of the complexity of the matrix and the trace levels at which the analytes must be detected. Capillary SFC has been shown to be a very effective technique for the analysis of trace levels of pesticides, especially when a selective detector such as the NPD is used (18). Figure 7 shows an example of the determination of selective carbamate pesticides in parsley. The sample was prepared by extraction of 1 gram of

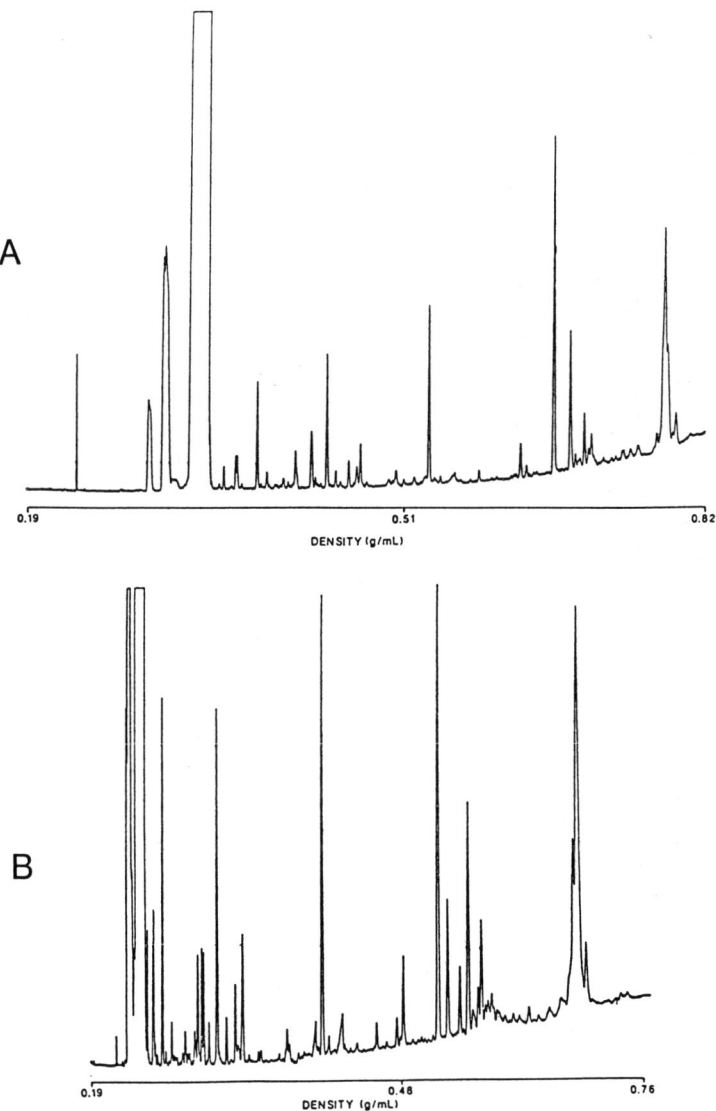

Figure 5. SFC chromatograms of cold pressed grapefruit oil. FID at 325°C, linear density programmed at 0.008 g/mL/min, 4 m x 50 μm i.d. SB-Biphenyl-30 column. Top chromatogram (A)--50°C oven temperature, bottom chromatogram (B)--100°C oven temperature.

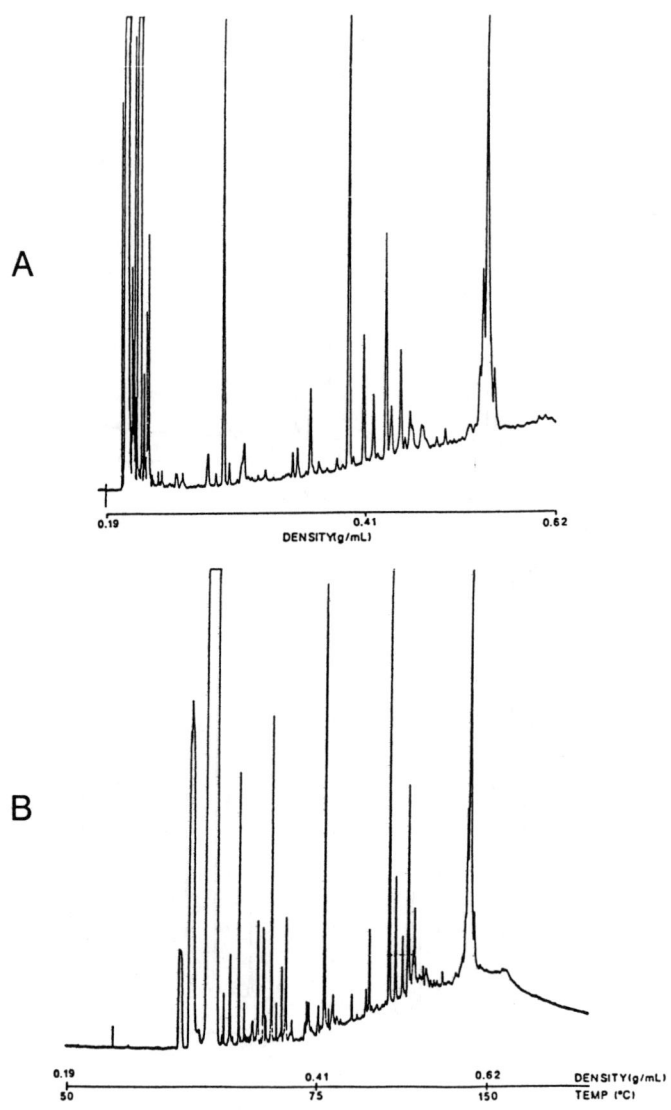

Figure 6. SFC chromatograms of cold pressed grapefruit oil. FID at 325°C, 4 m x 50 μm i.d. SB-Biphenyl-30 column. Top chromatogram (A)--150°C oven temperature programmed at 0.008 g/mL/min, bottom chromatogram (B)--simultaneous linear density/temperature programmed at 0.008 g/mL/min and 3°C/min.

parsley with 10 mL of acetone. The sample was concentrated to 0.5 mL, filtered and then analyzed. The pesticides were present at approximately 2 ng/g of parsley. Clearly, capillary SFC can be effectively used for the trace analysis of pesticide contaminants in some food stuffs.

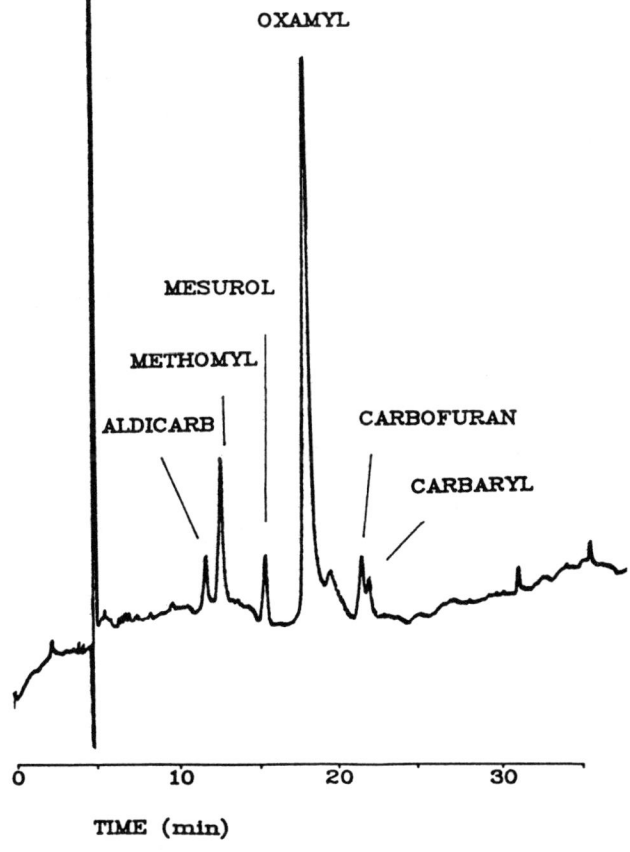

Figure 7. SFC chromatogram of carbamate pesticides in parsley. CO_2 at 100°C, 10 m x 50 μm i.d. SB-Methyl-100 column, NPD at 325°C, asymptotic density programmed.

Summary

The data presented here show that capillary SFC can be used to analyze food components such as fats, flavors and fragrances. The high chromatographic efficiency possible with capillary columns and the use of an FID make this technique an attractive alternative to HPLC or GC for some applications. When selective detectors are used, such as the NPD, SFC can also be used for the analysis of food contaminants such as pesticides.

Acknowledgments

Special thanks to A.A. Murakami of Miller Brewing for the hops extract sample and to K.C. Ting of the California State Department of Food and Agriculture for the parsley sample.

Literature Cited

1. Randall, L.G. Sep. Sci. Tech. 1982, 17, 1-118.
2. Klesper, E.; Corwin, A.H.; Turner, D.A. J. Org. Chem. 1962, 27, 700-701.
3. Novotny, M.; Springston, S.R.; Peaden, P.A.; Fjeldsted, J.C.; Lee, M.L. Anal. Chem. 1981, 53, 407 A-414A.
4. Chester, T.L. J. Chromatogr. 1984, 299, 424-431.
5. White, C.M.; Houck, R.K. J. High Resoln. Chromatogr. Chromatogr. Commun. 1985, 8, 293-296.
6. Chester, T.L.; Innis, D.P.; Owens, G.D. Anal. Chem. 1985, 2243-2247.
7. Chester, T.L.; Innis, D.P. J. High Resoln. Chromatogr. Chromatogr. Commun. 1986, 9, 209-212.
8. Chester, T.L.; Innis, D.P. J. High Resoln. Chromatogr. Chromatogr. Commun. 1986, 9, 178-181.
9. Copella, S.J.; Barton, P. in Supercritical Fluids: Chemical and Engineering Principles and Applications American Chemical Society, Washington, DC, 1987, ACS Symposium Series 329, p. 202.
10. Proot, M.; Sandra, P.; Geeraert, E. J. High Resoln. Chromatogr. Chromatogr. Commun. 1986, 9, 189-192.
11. Grob, Jr., K.; Neukom, H.P.; Battaglia, R. J. Amer. Oil. Chem. Soc. 1980, 282.
12. Mares, P.; Skorepa, J.; Sindelkove, E.; Turzicka, E. J. Chromatogr. 1983, 273, 172-180.
13. Colin, H.; Krstulovic, A.; Excoffier, J. L.; Guiochon, G. Liquid Chromatography Literature; Wiley: New York, 1984.
14. Stolyhwo, A.; Colin, H.; Guiochon, G. Anal. Chem. 1985, 57, 1342-1354.
15. Fisher, C.; Kalsec, Inc.; Personal Communication.
16. Murakami, A.A.; Miller Brewing; Personal Communication.
17. O'Shea, M.T.; Campbell Soup Company; Personal Communication.
18. Later, D.W.; Richter, B.E.; Felix, W.D.; Andersen, M.R.; Knowles, D.E. Amer. Lab 1986, 18(8), 108-115.

RECEIVED October 9, 1987

Chapter 11

Capillary Supercritical Fluid Chromatography–Mass Spectrometry

Practical Considerations and Applications

G. D. Owens, L. J. Burkes, J. D. Pinkston, T. Keough, J. R. Simms, and M. P. Lacey

Miami Valley Laboratories, The Procter & Gamble Company, Cincinnati, OH 45239-8707

> The combination of capillary supercritical-fluid chromatography with mass spectrometry promises to be an important new tool. It can be used where solutes can not be examined by GC-MS (due to thermal sensitivity and/or volatility). It also complements HPLC-MS by providing higher chromatographic efficiency and easier solvent elimination. This work focuses on mass spectrometer interface design, tuning techniques and several practical considerations encountered when performing SFC-MS determinations. Examples of both food and drug-related applications are presented.

Supercritical fluid mobile phases offer the chromatographer a unique combination of characteristics -- solvating properties similar to a liquid, and solute diffusivities intermediate between a gas and a liquid. In the 1960's, supercritical fluid mobile phases were used to separate thermally labile compounds (1) and then proposed as a means of extending the molecular weight range of gas chromatography (GC) (2,3). However, the potential of supercritical fluid chromatography (SFC) has not been extensively explored until recently. This revival of interest can be attributed to the development of fused-silica capillary columns with nonextractable stationary phases (4), the availability of commercial instrumentation (5), and the realization that SFC is uniquely able to solve a class of separations problems that falls between the capabilities of GC and high performance liquid chromatography (HPLC) (6).

The combination of a chromatographic technique with mass spectrometry (MS) provides a powerful analytical tool. This is evidenced by the dominance of GC-MS as the method of choice in volatile mixture analysis. The difficult coupling of HPLC with mass spectrometry has been pursued over the past two decades. The most promising LC-MS interface uses "thermospray" ionization (7) and is most applicable to polar compounds.

The idea of combining supercritical fluid introduction (SFI) or SFC with mass spectrometry is not new. The introduction of a supercritical fluid into a molecular beam instrument was proposed in the late sixties (8) and examined throughout the seventies (9-11). This work did not lead to a practical SFC-MS combination because of the complexity of the instrumentation and poor sensitivity. In the past four years, R. D. Smith and coworkers have reported progress in the development and application of capillary SFC-MS (12-18). The groups led by Henion (19,20) and Voorhees (21,22) have also added to this applications base. This body of work suggests that SFC-MS should be uniquely suited to the examination of non-polar compounds that are not volatile enough to be examined by GC-MS or are thermally labile.

In this report we summarize some of the practical considerations of coupling a capillary SFC instrument with a quadrupole mass spectrometer. Our initial results from food and drug-related examples demonstrate some of the promise this new technology holds.

EXPERIMENTAL

SFC Equipment. Our SFC-MS instrumentation is shown in Figure 1. The SFC part of this configuration is on the left side of the figure and has been described elsewhere (23). Liquid carbon dioxide, the mobile phase (SFC grade, Scott Specialty Gases, Plumbsteadville, PA), is introduced directly into a computer-controlled syringe pump. The pressure-programmed pump supplies the mobile phase to a 0.1-uL HPLC injector valve, IV, which is situated outside the chromatographic oven. Room-temperature, splitting injection occurs at the exit of the injection valve. A fraction of the analyte enters a 50-um i.d. retention gap, RG, at the split point. The retention gap is housed inside a 0.5-mm i.d. stainless steel tube. The majority of the analyte flows around the outside of the retention gap and into a stainless steel tee, T. A short length of 14-um i.d. fused-silica tubing, V, is connected to this tee. It serves as a vent for the majority of the analyte, and its length controls the split ratio.

After passing through the retention gap, a portion of the analyte is focused at the head of the analytical column, C. Two different fused-silica columns (J&W Scientific, Inc., Rancho Cordova, CA) were used depending on the application; a 10 m X 50 um i.d. X 0.2 um film thickness DB-1 column, or a 10 m X 50 um i.d. X 0.1 um film thickness DB-17 column.

MS Interface Probe. After separation on the capillary column, analytes are introduced directly into the mass spectrometer source via the interface probe shown in Figure 2. The heart of the probe is a 60 cm length of uncoated, 25 um i.d., fused-silica tubing (25VS-025ID, Scientific Glass Engineering, Inc., Austin, TX, USA). The end of this tube was tapered, sized, and cut to form a restrictor, R, in the manner previously described (23). The tapered end narrows to an aperture of approximately 3-5 um diameter over a length of 2 to 4 cm. This restrictor is housed inside a 50 cm length of 0.25 mm i.d. X 1.59 mm o.d. stainless steel tube, IT (3005, Alltech Assoc., Deerfield, IL, USA).

The inner tube, IT, is divided into two heated regions. A 2-cm length of IT near the tip of the probe has been wrapped with a layer of glass tape and a 30-cm length of 4Ω/ft nichrome wire (Pelican Wire Co., Naples, FL, USA) to form a tip heater, TH. This heater is

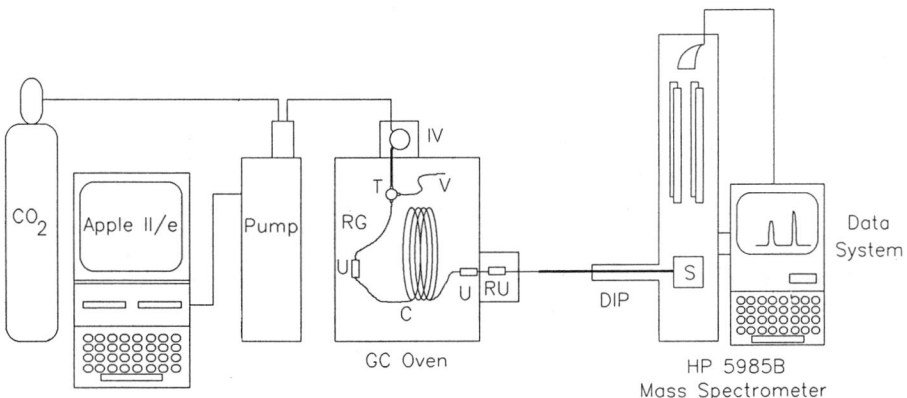

Figure 1. SFC-MS instrumentation.

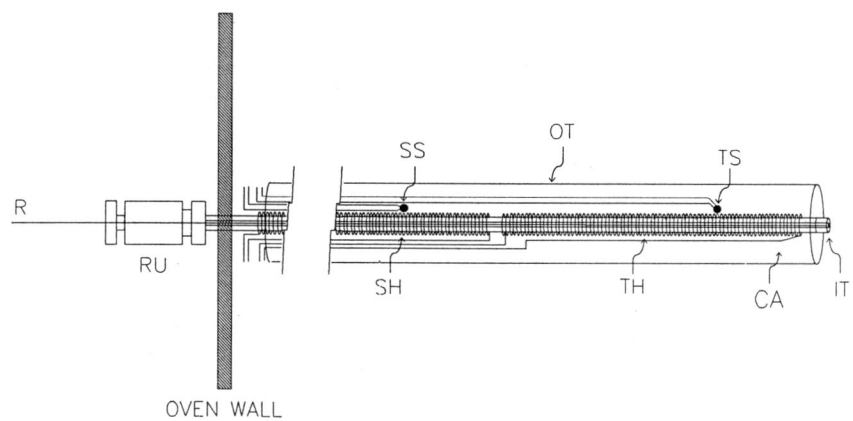

Figure 2. SFC-MS interface probe.

connected to a 22-W DC power supply (PAT 15-1.5, KEPCO, Flushing, NY, USA) by 22-guage, teflon-insluated leads which have been silver soldered to the nichrome wire. The temperature of the tip heater is measured by a 30 guage, iron-constantan thermocouple, TS (Omega Engineering, Stamford, CN, USA). This thermocouple and the entire tip heater are potted in a high-temperature ceramic adhesive, CA (AREMCO products Inc, Ossining, NY, USA).

The remaining 45 cm of IT is wrapped with glass tape and 1.6Ω/ft nichrome wire (Pelican) to form a stem heater, SH. The temperature of the stem heater is measured by a second, identical, iron-constantan thermocouple, SS. This heater is powered by a variable AC powerstat (10B, Superior Electric Company, Bristol, CN, USA). The entire stem heater is wrapped with a layer of glass tape. The oven-end of IT is coupled to a reducing union, RU (ZRU1.5, Valco). This union forms a vacuum seal between the restrictor and the inner tube.

The outer diameter of the interface probe and the shape of its tip are dictated by the mass spectrometer's direct insertion port and source inlet. The probe's outer tube, OT, is a 40 cm length of 6.4 mm o.d. X 4.8 mm i.d. stainless steel tubing (Microgroup Inc., Medway, MA, USA). A cylindrical bushing was custom machined from 316 stainless steel and silver soldered to both the inner tube, IT, and the outer tube, OT, to form a vacuum seal.

<u>MS Equipment</u>. The quadrupole mass spectrometer (Model 5985B, Hewlett-Packard) was operated in the chemical ionization (CI) mode. Various reagent gases were used, including isobutane, methane, methanol and ammonia (reagent grade or better, Matheson Gas Products, Secaucus, NJ). The analyzer had a mass range of 1000 Da. Unless specified otherwise, the MS ion-source temperature was 200°C, the electron multiplier was set to its maximum value of 3000 volts, and the scan rate was 6.66 sec/decade from m/z 200-1000. A cryopump (Cryo-miser IC-51, Torr Vacuum Products, Inc.) was used, when noted, to reduce the pressure in the source manifold. The partial pressure of CO_2 in the ion-source manifold, at an SFC pump pressure of 200 atm, was typically 1.0-1.5 X 10^{-4} torr. The CI reagent gas was added so that the manifold pressure increased by an indicated 1 X 10^{-4} torr. It should be noted that the response of the ionization gauge was not specifically calibrated for CO_2 or the various CI reagent gases. Under the conditions just described, the actual ion source pressure, measured with a thermocouple gauge, was 0.5 to 1.5 torr.

<u>Sample Preparation</u>. Triton-X 100 (Packard Instrument Company, Downers Grove, IL,USA) was dissolved in methylene chloride to a concentration of 25 mg/mL. Benzoyl peroxide (Aldrich, Milwaukee, WI) was dissolved to a concentration of 10 ug/uL in dry methylene chloride and stored at -20°C. Cholesterol (Lot 23F-7080, Sigma Chemical Co., St. Louis, MO) was dissolved in methylene chloride to a concentration of 1 ug/uL. A sample of Tergitol nonionic surfactant-9, USP name: Nonoxynol-9, (Union Carbide, S. Charleston, WV) was agitated, and an aliquot was taken to prepare a 3% (w/v) solution in methylene chloride. Samples of two different lots of food-grade siloxanes (DC200- 500 cSt, Dow Corning Corp., Midland, MI) were prepared with a concentration of 8% w/w in methylene chloride.

A 10% w/w solution of a lower viscosity polydimethylsiloxane (DC200-20 cSt, Contour Chemical Co., North Reading, MA) was also prepared in methylene chloride.

Trimethylsilyl (TMS) derivatives of oligomers of polyethylene glycol (PEG) 600 (Fisher Scientific, Springfield, NJ) were prepared according to the following derivatization procedure. One to fifty mg of sample was weighed into a one dram vial. The sample was then mixed with 0.2 mL pyridine, and 0.2 mL of a 5:1 mixture of trimethylsilylimidazol (TMSI) : N,O bis(trimethylsilyl)-tri-fluoroacetamide (BSTFA), (Supelco, Inc. Bellefonte, PA). This mixture was shaken and heated at 80°C for one hour. After cooling, 0.6 mL of methylene chloride was added. The final volume was 1.0 mL.

Methyl arachidate was made by methylation of arachidic acid. Approximately 1-2 mg of arachidic acid (product Number 10930, Fluka Chemical Corp., Ronkonkoma, NY) was reacted at room temperature for two minutes with 200 uL of distilled diethyl ether saturated with diazomethane (Aldrich Chemical Co. Inc., Milwaukee, WI) and 25 uL of methanol. The mixture was evaporated, and the resulting ester was diluted in hexane to produce a set of solutions with concentrations ranging from 1.2 ng/uL to 1.2 mg/uL.

RESULTS AND DISCUSSION

Several experimental variables have an important influence on the SFC-MS experiment using this equipment. These variables include the size of the restrictor orifice, the tuning of the mass spectrometer, the temperature of the MS-interface probe tip, and the pressure in the ion source manifold. All of these variables as well as the class and molecular weight of the analyte influence the sensitivity of the technique.

Restrictor Orifice Size. When splitting injection is used, a number of parameters vary as the restrictor orifice size is changed. If the splitter vent restrictor is held constant, then mobile phase velocity, sample loading on column (extent of column overload), mass of analyte per unit time to the detector, and resistance to plugging all increase with increasing restrictor aperture size. Conversely, chromatographic efficiency decreases with increasing restrictor orifice size and mobile phase velocity. Hence, the choice of the proper restrictor aperture for SFC-MS involves a trade-off between sensitivity and chromatographic efficiency. In this work, a restrictor aperture of 3-5 um was used because it provided an acceptable compromise between these two characteristics. This aperture size is larger than that used commonly with the FID.

Mass Spectrometer Tuning. A three-step process was used to tune the quadrupole mass spectrometer prior to its use as a detector for SFC. In the first step, perfluorotributylamine (Pfaltz Bauer Inc., Stamford, Conn.) was ionized by electron ionization and used to calibrate the mass axis. In the second step, methane was introduced into the CI source and the reagent ion profiles were optimized. In the third step, the mass resolution was adjusted for improved sensitivity. This was accomplished by introducing a volatile brominated compound, such as 2-bromopentane, into the CI source. The mass spectrometer's resolving power was reduced such that the peaks

of the protonated molecule isotopic doublet (separated by 2 Da) were
separated by a 50% valley. Operating the mass spectrometer at this
decreased resolving power provided the expected ten-fold sensitivity
enhancement compared to operation at unit resolving power (24).

SFC-MS Interface Tip Temperature. Figure 3 shows the effect of
increasing MS interface probe-tip temperature on the separation of
the oligomers in the surfactant Triton X-100. At probe tip
temperatures below 300°C, the restrictor consistently

$$CH_3-\underset{\underset{CH_3}{|}}{\overset{\overset{CH_3}{|}}{C}}-CH_2-\underset{\underset{CH_3}{|}}{\overset{\overset{CH_3}{|}}{C}}-\underset{}{\bigcirc}-O[CH_2CH_2O]_{\overline{n}}-H$$

Triton X-100

plugged before a separation could be completed. Figures 3a and 3b
indicate that a plateau in the reconstructed total-ion-current (RTIC)
response exists for this compound between 300°C and 350°C. The
response at 400°C, Figure 3c, shows a reduction in both the signal
and the background current and a pronounced shift to longer retention
times. The shift to longer retention times and lower ion current at
higher temperature is presumably due to an increase in moble phase
viscosity in the restrictor tip. This viscosity increase causes a
decrease in both the mobile phase velocity and column loading. The
mass spectra of the oligomers at 400°C showed no evidence of
thermal decomposition. This example demonstrates that a number of
factors change with interface-tip temperature. We chose a
tip-temperature range of 325-350°C for the separation of
thermally-stable compounds similar to Triton X-100. This value
changes and must be empirically established for each type of compound
examined.

Routine operation of the mass spectrometer interface tip at high
temperature might cause the decomposition of thermally-labile
materials. Historically, cholesterol has been used to test the
thermal activity of GC-MS interfaces and jet separators. It is very
sensitive to dehydration across the 3-4 bond upon contact with
"active sites" in the system, yielding a species with molecular
weight 368 Da (25).

Cholesterol. MW 386 Da

Cholesterol was examined by both SFC-MS and GC-MS under identical
isobutane CI conditions. SFC-MS analyses were made at a probe-tip
temperature of 350°C. GC-MS analyses were made with an interface

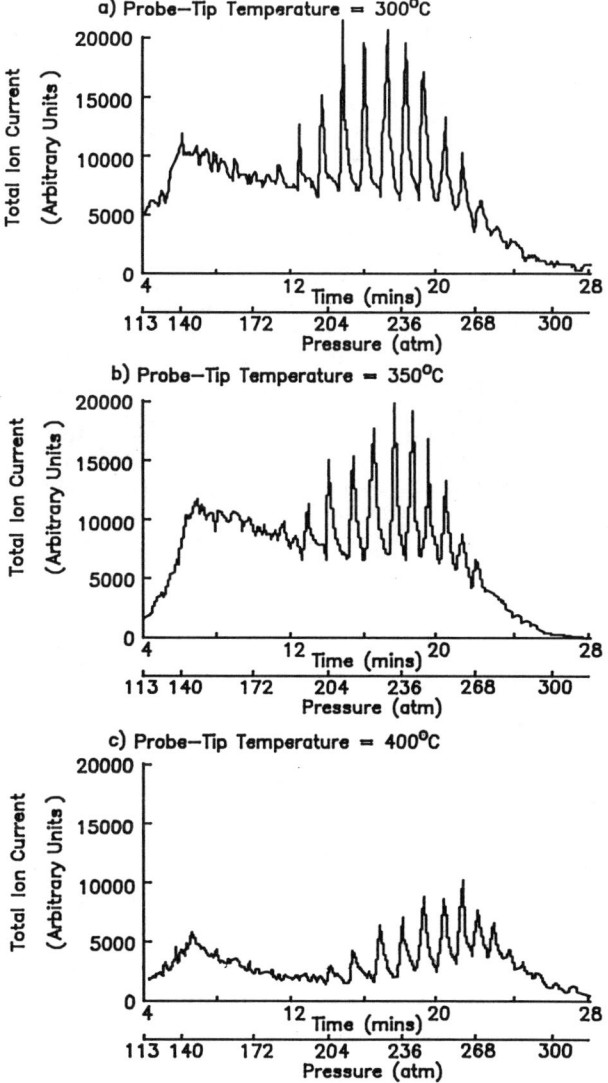

Figure 3. SFC-MS response vs. probe tip temperature for Triton X-100. Conditions: 10 m X 50 um i.d. X 0.1 um film DB-17 column, column temperature = 90°C, probe stem temperature = 90°C, 0.1 uL of a 25 mg/ml Triton X-100 in methylene chloride solution split injected, split ratio ~1:2, pressure program = 100 atm for 3 min, ramp to 140 atm in 3 min, ramp to 325 atm in 23 min, methane CI, MS source temperature = 200°C.

temperature of 300°C. During the GC-MS runs, CO_2 was introduced through the SFC probe into the CI volume. SFC-MS analyses were performed with the introduction of helium into the ion source through the GC transfer line. The isobutane CI spectra of cholesterol obtained by GC-MS and SFC-MS under these ionization conditions were virtually identical. In both cases the m/z 369 peak was the base peak, and the m/z 386 peak had a relative intensity of approximately 10%. This demonstrates that the SFC-MS interface does not contribute to "active site" thermal degradation to any greater extent than does the widely accepted fused-silica GC-MS interface.

SFC has been used for separations of thermally-labile compounds (26) due to its mild elution conditions as compared to those of GC. We examined benzoyl peroxide as a second, more sensitive, test of our interface's performance with thermally-labile compounds. Benzoyl peroxide is a relatively stable peroxide, but it is still very thermally labile (27) and difficult to analyze by GC, as are most peroxy-compounds (28). The peroxide linkage is prone to cleavage, and common decomposition products are known (27). The SFC-FID analysis of benzoyl peroxide provides a single, sharp peak, if the chromatographic oven is maintained at temperatures < 100°C. This suggests that the peroxide is stable during the SFC analysis, or that it is converted to a stable product upon or before injection.

A solution of benzoyl peroxide in methylene chloride was analyzed by SFC-MS using both isobutane and ammonia CI. The chromatographic oven and probe stem where held at 60°C, the probe tip was held at 250°C, and the mass spectrometer source was held at 200°C. Under these conditions, a single peak in the RTIC is observed. However, the major species observed in the mass spectra are attributable to the protonated molecules of known thermal decomposition products: phenolic ester of benzoic acid, benzoic acid, and phenol. Only a small fraction (<5% relative intensity) of the total ion current corresponded to the intact, protonated benzoyl peroxide. This mixture of products and the single chromatographic peak suggest that the benzoyl peroxide survived the SFC separation intact. Decomposition is probably occurring in the source or in the interface. Additional experiments, in which both the source and SFC-MS interface temperatures are varied, will be required to establish the site of decomposition. It is clear, however, that the application of SFC-MS to thermally-labile materials will ultimately be limited, not by the SFC, but by the interface-detector combination.

<u>Ion Source Pressure</u>. During pressure-programmed SFC-MS runs, we have observed an apparent decrease in signal-to-noise ratio (S/N) as the pressure in the source manifold increases. Data from the manufacturer of the mass spectrometer suggest that an eight-fold decrease in relative sensitivity should be expected as the pressure in the source manifold region increases from 1×10^{-4} to 5×10^{-4} torr (29). Figure 4 illustrates this effect. This figure shows the RTIC vs. time profiles for two SFC separations of the silylated oligomers of polyethylene glycol (PEG) 600. Both analyses were conducted with equal amounts of PEG injected. The upper trace shows the response when cryopumping was used during the entire run and the source manifold pressure remained below 8×10^{-5} torr. The lower trace shows the response for an identical run without cryopumping

where the source pressure rose to over 3×10^{-4} torr as the later peaks were eluting. These results show that reducing the source manifold pressure improves the S/N by at least a factor of two. The S/N improves by almost a factor of 4 for the higher molecular weight species which elute at increased mobile phase pressure. Sensitivity losses at higher pressure are apparently due to the loss of ions by scattering, and other collisional processes, as they travel between the exit of the CI volume and the entrance of the analyzer.

Figure 5 shows the RTIC and selected mass chromatograms for the derivatized oligomers of PEG 600 when cryopumping is employed. Protonated molecular species were observed for compounds with masses up to the 1000 Da mass limit of this mass spectrometer. The species at m/z 999 has 19 ethylene glycol repeating units.

Sensitivity and Chromatographic Efficiency. We used methyl arachidate to evaluate the sensitivity of the SFC-MS combination using the HP quadrupole instrument. A splitless injection of 1.2 ng of methyl arachidate in dichloromethane produced a peak in the mass chromatogram of m/z 327, $[M+H]^+$. The S/N was roughly 5 at the expected retention time. The instrument was operated in the isobutane CI mode at unit resolution and was scanned from m/z 120 to m/z 500 in 1 s. This level of sensitivity is comparable to that observed by other workers performing SFC-MS with electron ionization (22). However, much better sensitivity values have been reported for chemical ionization of both biphenyl compounds (15) and trichothecene mycotoxins (18). The reason for this difference in CI mode sensitivities is not presently understood. Certainly instrumental configuration, method of ionization, nature and molecular weight of the species in question, and the transmission/detection efficiency of the mass spectrometer all affect sensitivity.

The chromatographic efficiency of the SFC-MS combination with the HP quadrupole as a detector is lower than what we commonly observe using SFC-FID. Chromatographic efficiency has been sacrificed to gain sensitivity. Larger restrictor apertures (which produce larger column flow velocities, smaller split ratios and more mass on column given a constant splitter vent restriction), and higher pressure program rates decrease the apparent efficiency and improve sensitivity. Smith (18) has discussed the use of very high pressure-program rates as a means of improving sensitivity. The complexity of the mixtures examined in this work precluded the use of this technique.

Application of SFC-MS to a Drug-Related Problem. Tergitol nonionic surfactant-9 (USP name: Nonoxynol-9; abbreviated here TNS-9) is an important active ingredient and surfactant in a number of drug formulations.

$$C_9H_{19}-\phenyl-O[CH_2CH_2O]_n^-H$$

Tergitol nonionic surfactant-9.

Figure 6 shows the SFC-FID chromatogram of TNS-9. It exhibits a series of similar multiplets, each consisting of 6 or 7 major peaks or shoulders. Spiking experiments, in which standards were injected

200   SUPERCRITICAL FLUID EXTRACTION AND CHROMATOGRAPHY

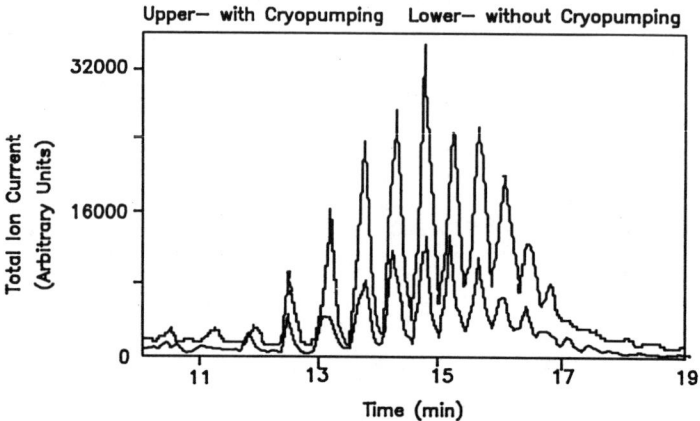

Figure 4. SFC-MS response vs. ion-source manifold pressure for
PEG 600. Conditions: split injection (~1:2) of 0.1 uL
of a 20 mg/mL solution of PEG 600 in methylene
chloride, 10 m X 50 um i.d. X 0.1 um film DB-17 column,
column temperature = 90°C, probe stem temperature =
90°C, probe tip temperature 350°C, pressure program
= 100 atm for 3 min, ramp to 140 atm in 3 min, ramp to
365 atm in 22 min, isobutane CI, MS source temperature
= 200°C.

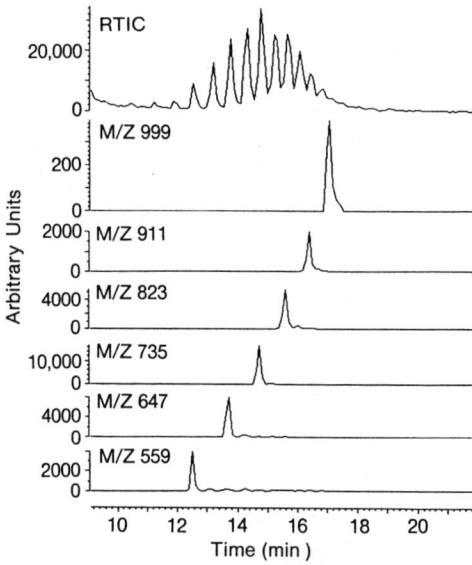

Figure 5. RTIC and selected mass chromatograms for PEG 600 using
cryopumping. Same operating conditions as Figure 4.

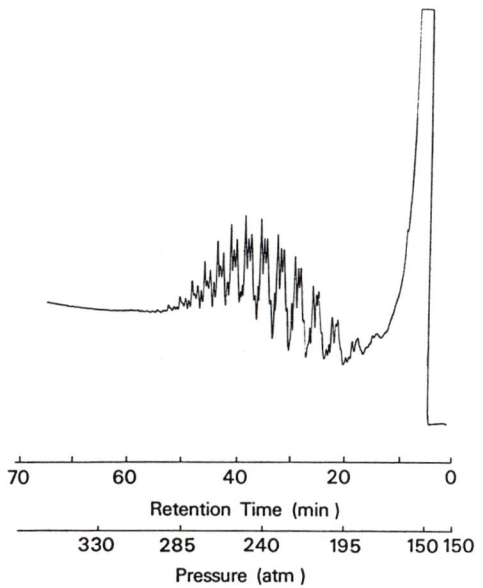

Figure 6. SFC-FID chromatogram of Tergitol Nonionic Surfactant-9. Conditions: 10 m X 50 um i.d. X 0.1 um film DB-17 column, column temperature = 100°C, detector temperature = 70°C, pressure program = 150 atm for 5 min, ramp to 360 atm at 3 atm/min.

with the unknown, suggested that each multiplet corresponds to a specific ethoxylate chain length (one value of n), and to a variety of isomers of an 8, 9 or 10 carbon side-chain.

To confirm our assignments, we ran TNS-9 by SFC-MS. Figure 7 shows the RTIC for the run as well as one mass spectrum from a peak in a typical multiplet. The RTIC resembles the FID trace. In spite of the fact that the chromatographic efficiency is much lower than in the SFC-FID run, one can still recognize the TNS-9 multiplets in the first half of the chromatogram. All spectra collected across a single multiplet exhibited the same protonated molecule. In addition to the protonated molecule, the base peak in all the spectra, several series of fragment ions, separated by 44 Da intervals, were observed. These results indicate that each multiplet corresponds to isomers of a particular ethoxylate chain length and of a hydrocarbon chain length of nine. The complexity of the multiplets is presumably due to hydrocarbon chain branching and/or in positional isomerism about the phenyl ring.

The mass spectrum shown in Figure 7 does not exhibit stable isotopic multiplets and the signal intensity is weak. The lack of stable isotopes is due to the decreased mass resolution. However, this example demonstrates that even low quality mass spectra can be helpful for the characterization of materials. The most important result for this experiment is that one SFC-MS analysis gave us better information on the nature of TNS-9 than did over a week of careful SFC-FID work.

Application of High-Mass SFC-MS to Food-Related Problems. The characterization of mixtures of food-grade polydimethylsiloxane (PDMS) oligomers is a difficult problem. The large number of oligomers requires that a high-resolution separation technique be used. However, the molecular weight range encountered is well beyond the range of any form of gas chromatography. A powerful characterization technique is badly needed because the performance of these materials in food applications varies widely from lot to lot, even when the viscosity (commonly used to characterize PDMS samples) is constant. We have used SFC-FID to examine samples from two different lots of a food-grade PDMS (30) that exhibited different product performance. "High-mass" SFC-MS has aided our understanding of the molecular weight range of these separations.

We have recently interfaced a capillary SFC instrument using splitless injection to a 3000 Da, VG quadrupole mass spectrometer (31). Figure 8 shows the RTIC and representative spectra collected from an SFC-MS run of a 20 cSt PDMS sample. All of the oligomers show $[M+NH_4]^+$ ammonium adduct ions and the assigned m/z values are within one unit of the expected centroid. From this SFC-MS run, we were able to obtain molecular weight information up to the 39th oligomer (ammonium adduct ion of m/z 2998) for the 20 cSt PDMS sample. The SFC-FID chromatogram for this material shows well-resolved peaks up to the 65th oligomer. Based on the SFC-MS results, we can assign the mass for the later eluting peaks. Comparison of the SFC-FID chromatograms from the 20 cSt PDMS and the 500 cSt food-grade materials allows us to estimate a molecular-weight range of at least 8500 for SFC separations of the higher-viscosity, food-grade materials.

High-mass SFC-MS has also been used to examine the oligomers in

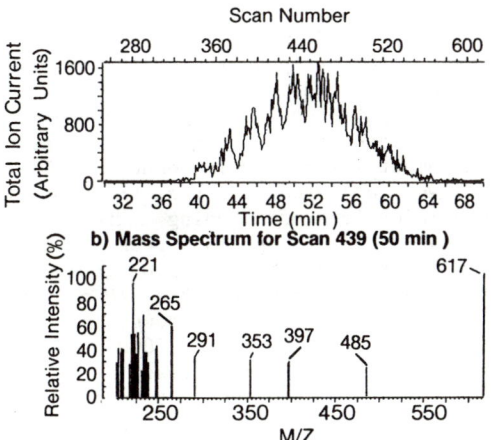

Figure 7. a) SFC-MS RTIC for Tergitol Nonionic Surfactant-9. Conditions: split injection (~1:2) of 0.1 uL of a 30 mg/mL solution of TNS-9 in methylene chloride, 10 m X 50 um i.d. X 0.1 um film DB-17 column, column temperature = 90°C, probe stem temperature = 90°C, probe tip temperature = 350°C, pressure program = 80 tm for 5 min, ramp to 360 atm in 80 min, isobutane CI, MS source temperature = 200°C.

b) A Selected Mass Spectra for the n=9 Tergitol Oligomer.

silylated corn sweeteners up to 3000 Da (31). The trimethylsilyl derivatives of oligosaccharides can be easily eluted with unmodified CO_2 mobile phase in capillary SFC over a wide molecular weight range (32). The oligomes elute as a series of irregularly spaced doublets. Figure 9 illustrates the RTIC chromatogram for the oligosaccharides containing two through seven units (G2-G7). The molecular weights of the oligosaccharides G8 and above were beyond the mass range of the mass spectrometer. Figure 9 also illustrates representative spectra collected during the NH_3 CI run. Each oligosaccharide displays a base peak within 1 mass unit of that expected for its ammonium adduct ion. The more intense spectra usually displayed fragment ions (m/z 773, 1152) related to the structure of the oligosaccharide. Analyzing this mixture by SFC-MS allowed us to strengthen our assignment of the doublets as anomeric forms of the oligosaccharide (33).

During the runs of high molecular weight oligomeric series such as the polydimethylsiloxanes and the oligosaccharides, the resolving power of the VG instrument was purposely reduced. The peak detection parameters were adjusted such that the isotopic multiplets corresponding to the major ions of a tuning mixture merged into single, centroided peaks. (The "high mass peak width" in the VG tuning parameters file was increased from 1.02 mass units to 1.20 mass units.) This dramatically improved the S/N of the measurements. The reduction in resolving power increased the signal intensity by a factor of roughly 10 at m/z 2000 as directly measured on the oscilloscope during the tuning process.

SFC-MS conditions were simulated while tuning the VG mass spectrometer by introducing a 3% solution of the 20 cSt PDMS in supercritical CO_2 into the ion source. SFI of the PDMS mixture provided oligomers which ranged in molecular weight from 200 daltons to well over 4000 daltons.

The sensitivity of this SFC-MS combination was evaluated with two standards: methyl arachidate (326 Da) and tristearin (890 Da). The instrument was tuned to "unit" resolution (10% valley definition) before the sensitivity measurements were made. Splitless injection of 800 pg of methyl arachidate produced a mass chromatogram of m/z 344, $[M+NH_4]^+$, with a S/N of approximately 4. The scan range was m/z 100 to 500 with a 1 s scan cycle time. Splitless injection of 12 ng of tristearin produced a mass chromatogram of m/z 908, $[M+NH_4]^+$, with a S/N of approximately 5. The scan range was from m/z 500 to 1000 with a scan cycle time of 1.1 s. The tristearin result is a "worst case" result. Logistical difficulties of combining instruments from two distant sites over a short period of time precluded optimization of S/N.

These results show for the first time that it is possible to generate and detect high-mass ions (> 1500 Da) after their parent compounds have been introduced into a mass spectrometer by SFC. Both of these examples, PDMS and silylated oligosaccharides, are mixtures of relatively volatile compounds. We have not succeeded in observing ions related to less volatile materials, like sucrose octaoleate, that require high SFC elution pressures (370 atm). Future SFC-MS work will examine the applicability of this technique to high mass species that are less volatile. The combination of this unique separations method and a high-mass quadrupole mass spectrometer will be a powerful analytical tool.

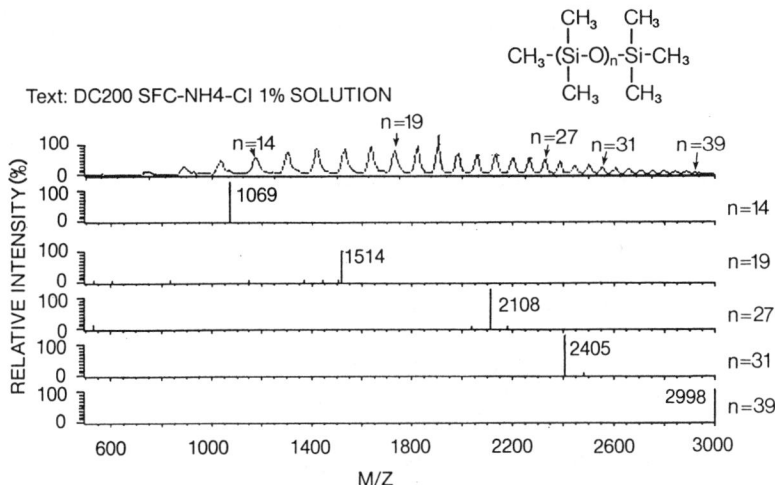

Figure 8. Reconstructed total-ion-current chromatogram (top) and selected mass spectra produced during NH_3 CI SFC-MS run of 20 cSt PDMS. (Adapted from Ref. 31. Copyright 1988 ACS.)

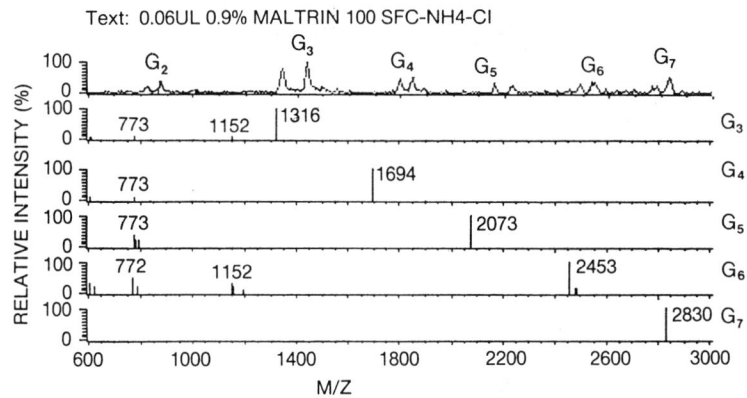

Figure 9. Reconstructed total-ion-current chromatogram and selected spectra from NH_3 CI SFC-MS run of derivatized Maltrin 100. (Adapted from Ref. 31. 1988 ACS.)

Literature Cited

1. Klesper, E.; Corwin, A. H.; Turner, D. A. J. Org. Chem. 1962, 7, 700-701.

2. Sie, S. T.;Van Beersum, W.; Rijnders,G. W. A. Sep. Sci. 1966, 1, 459-490.

3. Myers, M. N.; Giddings, J. C. Sep. Sci. 1966, 1, 761-776.

4. Novotny, M.; Springston, S. R.; Peaden, P. A.; Fjeldsted, J. C. Lee, M. L. Anal. Chem. 1981, 53, 407A-414A.

5. Gere, D. R. Science 1983, 222, 253.

6. Chester, T. L. J. Chrom. Sci. 1986, 24, 226-229.

7. Vestel, M. L. Mass Spectrom. Rev. 1983, 2, 447-480.

8. Milne, T. A. Int. J. Mass Spectrom. Ion Phys. 1969, 3, 153-155.

9. Giddings, J. C.; Myers, M. N.; Wahrhaftig, A. L. Int. J. Mass Spectrom. Ion Phys. 1970, 4, 9-20.

10. Randall, L. G.; Wahrhaftig, A. L. Anal. Chem. 1978, 50, 1705-1707.

11. Randall, L. G.; Wahrhaftig, A. L. Rev. Sci. Instrum. 1981, 52, 1283-1295.

12. Smith, R. D.; Felix, W. D.; Fjeldsted, J. C.; Lee, M. L. Anal. Chem. 1982, 54, 1883-1885.

13. Smith, R. D.; Fjeldsted, J. C.; Lee, M. L. J. Chromatogr. 1982, 247, 231-243.

14. Smith, R. D.; Udseth, H. R. Anal. Chem. 1983, 55, 2266-2272.

15. Smith, R. D.; Udseth, H. R.; Wright, B. W.; Kalinoski, H. T. Anal. Chem. 1984, 56, 2476-2480.

16. Wright, B. W.; Udseth, H. R.; Smith, R. D. J. Chrom. 1984, 314, 253-262.

17. Smith, R. D.; Udseth, H. R.; Kalinoski, H. T. Anal. Chem. 1984, 56, 2971-2973.

18. Smith, R. D.; Udseth, H. R.; Wright, B. W. J. Chrom. Sci. 1985, 23, 192-199.

19. Crowther, J. B.; Henion, J. D. Anal. Chem. 1985, 57, 2711-16.

20. Lee, E. D.; Henion, J. D. HRC&CC 1986, 9, 172-174.

21. Holzer, G.; Deluca, S.; Voorhees, K. J. HRC&CC 1985, **8**, 528-531.

22. Zaugg, S. D.; Deluca, S. J.; Holzer, G. U.; Voorhees, K. J. HRC&CC 1987, 10, 100-101.

23. T. L. Chester, D. P. Innis, and G. D. Owens, Anal. Chem. 1987, **57**, 2243-2247.

24. Dawson, P. H. Quadrupole Mass Spectrometry and Its Applications; Elsevier Scientific: Amsterdam, 1976; p 141.

25. Krueger, P. M.; McCloskey, J. A. Anal. Chem. 1969, **41**, 1930.

26. Markides, K. E.; Fields, S. M.; Lee, M. L. J. Chromatogr. 1986, **24**, 254.

27. Richter, F. Beilsteins Handbuch Der Organischen Chemie; Springer-Verlag: Berlin, 1949; EII9, H9, 158.

28. Cairns, G. T.; Ruiz Diaz, R.; Selby, K.; Waddington, D. J. J. Chromatogr. 1975, **103**, 381.

29. Data courtesy of Paul Goodley, Hewlett Packard Co., Palo Alto, CA.

30. Chester, T. L.; Burkes, L. J.; Delaney, T. E.; Innis, D. P.; Owens, G. D.; Pinkston, J. D. ACS Symposium Series; this volume.

31. Pinkston, J. D.; Owens, G. D.; Millington, D. S.; Burkes, L. J.; Delaney, T. E.; Maltby, D. A., in press.

32. Chester, T. L.; Innis, D. P. J. High Resolu. Chromatogr. Chromatogr. Commun. 1986, **9**, 209.

33. Chester, T. L.; Pinkston, J. D.; Owens, G. D. Carbohydr. Res., in preparation.

RECEIVED November 13, 1987

Chapter 12

Supercritical Fluid Chromatography–Mass Spectrometry of Carotenoid Pigments

Nelson M. Frew[1], Carl G. Johnson[1], and Richard H. Bromund[2]

[1]Department of Chemistry, Woods Hole Oceanographic Institution, Woods Hole, MA 02543
[2]Department of Chemistry, College of Wooster, Wooster, OH 44691

>   Combined supercritical fluid chromatography–mass spectrometry is shown to be a useful new tool for the separation and identification of carotenoids, relatively involatile, labile pigments which contain multiple functional groups spanning a range of polarities. The most promising stationary phases for capillary SFC of complex natural mixtures of carotenoids are the cyanopropylpolysiloxanes and polyethylene glycols. The extremely mild ionization conditions which prevail using supercritical CO_2 as the mobile phase, produce superior quality mass spectra for fragile carotenoids such as fucoxanthin and its derivatives, as compared with earlier in-beam desorption CI techniques. The CI-CH_4 fragmentation of many other carotenoids under these conditions is minimal; the simplicity of their spectra may be advantageous in determining low level distributions using molecular ion abundances.

The separation and structural characterization of complex mixtures of natural products is an important problem in food and agricultural chemistry, pharmaceutical research and environmental and geochemical studies. Gas chromatography (GC) and high performance liquid chromatography (HPLC), now well-established methods, have largely supplanted earlier techniques, including thin-layer chromatography and open-column liquid chromatography, for reasons of chromatographic efficiency and speed of analysis. The use of combined gas chromatography–mass spectrometry (GC-MS) has been particularly valuable in the analysis of complex mixtures, and more than a decade of research on combined HPLC-MS has resulted in interfacing techniques such as thermospray ionization and electrospray ionization which show similar promise.

The introduction of open-tubular columns and improved hardware for supercritical fluid chromatography (SFC) has renewed interest in the use of supercritical fluids as mobile phases for chromatographic separations (1-3). This new technique supplements the separation capabilities of both GC and HPLC for two reasons: (1) the ability to deal with less volatile, thermolabile and high molecular weight materials not generally amenable to GC analysis, but with higher chromatographic efficiencies than are attainable in HPLC and (2) the greater flexibility in available detection modes, including UV-visible, fluorescence, flame ionization, thermionic, Fourier transform infrared and mass spectrometric detection. The potential for combining SFC with the latter of these detectors, the mass spectrometer (MS), is an important advantage because of the universality and specificity of MS and its ability to supply explicit structural information. Several of the common MS ionization modes, including electron ionization (EI), chemical ionization (CI) and charge-exchange (CE) have been shown to be compatible with on-line SFC (4-12). Varying degrees of hardware modification are required, but are generally much less extensive than required in HPLC-MS, where the mass spectrometer must cope with high flow rates of polar liquids.

The increasing prominence of SFC is reflected in the many recently-published applications, primarily involving industrial uses, including analysis of synthetic surfactants, polymers, food and cosmetic formulations and petroleum distillates (13-19). Relatively little work has been published on SFC of natural products, particularly of polar materials with molecular weights greater than 400 daltons (20-23). The choice of polar supercritical fluids is relatively limited in practice, since critical temperatures and chemical reactivity increase dramatically with polarity. Some progress has been made with the use of polar modifiers, but most investigations of mixed fluids have been confined to studies of elution order effects with relatively non-polar or low molecular weight polar solutes (24-29). The extent to which SFC will be useful for relatively polar materials (and thus a strong competitor with HPLC) is still an open question and a matter for much future research.

We have selected a class of labile polar lipids, the carotenoids, with which to explore the potential of SFC-MS. We are specifically interested in these compounds as biological source markers and as model indicators of the fate of labile organic matter which is produced in the surface ocean and is subject to various biological and chemical degradation processes. However, our findings should be of general interest to others working to apply SFC-MS in other areas of natural product chemistry.

The carotenoids (carotenes and xanthophylls) are widely distributed in both photosynthetic and non-photosynthetic organisms, functioning as accessory light receptors assisting in the transfer of energy to the chlorophylls and acting as antioxidants which protect the chlorophylls from photooxidation during photosynthesis. The carotenoids are tetraterpenoids (Figure 1), each consisting of a conjugated polyene backbone often terminated at each end with six-membered rings of varying unsaturation. The terminal rings (as well as the backbone) are typically substituted with various polar functional groups, including hydroxyl, methoxy, keto, acetyl

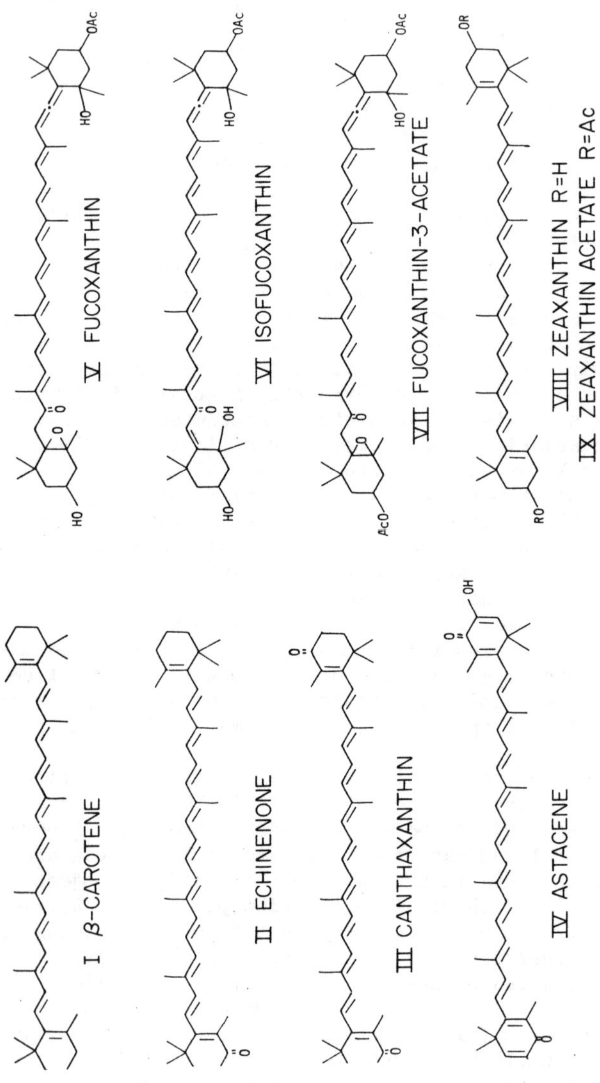

Figure 1. Structures of selected carotenoid pigments studied in this work.

and epoxy groups. The highly reactive nature of these compounds which makes them attactive as indicators of short-term degradative pathways also makes their isolation and mass spectral characterization difficult, particularly at low levels (< 100 ng). Carotenoids are not generally amenable to gas chromatographic analysis due to their thermolability and low volatility (35, 42). The analysis of carotenoids currently involves separation by HPLC and identification of the purified compounds by absorption spectroscopy and direct insertion probe mass spectrometry (30). The electron ionization, chemical ionization and field desorption fragmentation of carotenoid pigments have been studied by a number of workers (31-33). Vetter et al. (31) have discussed in detail the problem of thermal degradation of carotenoids and the observation that carotenoid EI fragmentation patterns, to a large extent, actually represent thermally degraded and isomerized carotenoids. Combined on-line chromatography-mass spectrometry (HPLC-MS or SFC-MS) has not been previously demonstrated for the carotenoids.

We have investigated the possibility of identifying carotenoid pigments using capillary supercritical fluid chromatography-mass spectrometry. In view of their lability, low-volatility, and polarity range, these compounds present an interesting test of the current capabilities of capillary SFC-MS. We demonstrate that SFC-MS can be implemented with relatively few modifications to a typical quadrupole GC-MS system and that useful chromatography and structural information may be obtained for these compounds using SFC.

Instrumentation and Experimental Methods

Two supercritical fluid chromatographs were used in this work. The first system, used for SFC alone, was constructed using a computer-controlled (Apple IIe) pressure/density programmable Lee Scientific Model 250 syringe pump and a Hewlett-Packard 5710 chromatograph modified to accept a Rheodyne 7520 injector (rotor internal loop volume, 0.2 μl). Columns were connected to the Rheodyne injector through a splitter tee either directly or indirectly using a short retention gap of uncoated capillary. Jet-type restrictors were fabricated directly on the columns by flame sealing the column end to form a sharp internal taper, then opening the seal by grinding to the desired orifice diameter of approximately 1 to 3 microns (34). When flame ionization detection (FID) was used, some flame instability was observed at high pressures unless the restrictor was positioned slightly below the tip of the FID jet. This necessitated heating the FID to temperatures of 300-350°C to avoid peak tailing.

A second system used for SFC-MS consisted of a Brownlee Model G Micropump, a Rheodyne 7520 injector (0.2 μl rotor; splitter tee) and a Carlo Erba 4160 gas chromatograph. The injector was chilled to 18°C using a refrigerated circulating bath. The chromatograph was directly coupled to a Finnigan 4500 quadrupole mass spectrometer (essentially without modification using the same electrically heated interface oven used for GC-MS operation) such that the column effluent entered the ion volume normal to the quadrupole axis. The column restrictor was positioned 0.5 mm back from the sample entrance hole. While electrical heating maintained the interface region at the same temperature as the chromatograph oven, no additional heat was supplied to the restrictor other than that from the source block, which was maintained at 100-120°C. Methane chemical ionization was achieved using a methane pressure of 0.5 torr (uncorrected). Analyzer pressures ranged from 3-6 x 10^{-5} torr. The electron energy was 130 eV. The electron multiplier was operated at 1.3 kV with the conversion dynodes at 3 kV. Preamplifier gain was 10^{-8} A/V. The spectrometer was tuned for unit resolution and calibrated in the CH_4-CI mode using perfluorotributylamine. Data acquisition was carried out with an INCOS 2300 data system. Data were acquired over the mass range of 400-800 daltons using two-second scans to enhance the signal-to-noise ratio.

Separations were carried out using 50 or 100 µm ID DB-5, DB-17, DB-225 and DB-WAX coated fused silica capillary columns (J&W Scientific, Rancho Cordova, CA) of 2 to 20 meters in length. SFC Grade carbon dioxide (Scott Specialty Gases, Plumbsteadville, PA) was used as the carrier fluid at 60-100°C with carrier linear flow velocities of 2-5 cm/sec. Underivatized carotenoid standards (isolated from natural sources by D. Repeta or obtained from Sigma Chemical Co., St. Louis, MO) were dissolved in methylene chloride to a concentration of 400-4000 ng/µl and injected using a 0.2 µl rotor and a split ratio in the range of 2-10:1.

Results and Discussion

Chromatography of Carotenoids. One of the earliest literature reports on SFC demonstrated the migration and separation of alpha- and beta-carotene in supercritical CO_2 (35). Enhanced solubility of these compounds in CO_2 was ascribed to formation of acceptor-donor complexes with the polarizable pi-electrons in the highly conjugated backbone. In the present work, all of the carotenoids shown in Figure 1, as well as several related isomers, are found to be sufficiently soluble in supercritical CO_2 to allow chromatographic elution from the stationary phases tested. We examined the chromatographic behavior of carotenoids on several types of stationary phases using simple mixtures spanning a range of polarities. In general, selectivities for the carotenoids on polysiloxane phases are limited. On DB-5 (5% phenyl, methyl-polysiloxane), relatively sharp chromatographic peaks are obtained, but the selectivity is inadequate to separate even simple mixtures. Useful separations are obtained only on moderately polar to polar stationary phases. Separation of several carotenoids on a 20 m x 100 µm ID DB-17 (50% phenyl, methyl-polysiloxane) column is illustrated in Figure 2A. The unusual elution behavior of fucoxanthin (V) is exemplified by its elution from DB-17 well before other carotenoids. This behavior is temperature sensitive and not exhibited by other polar carotenoids. DB-225 (50% cyanopropylmethyl, 50% methylphenyl-polysiloxane) exhibits similar moderate selectivity (Figure 2B), but still does not provide sufficient resolution of the more polar carotenoids. The highest

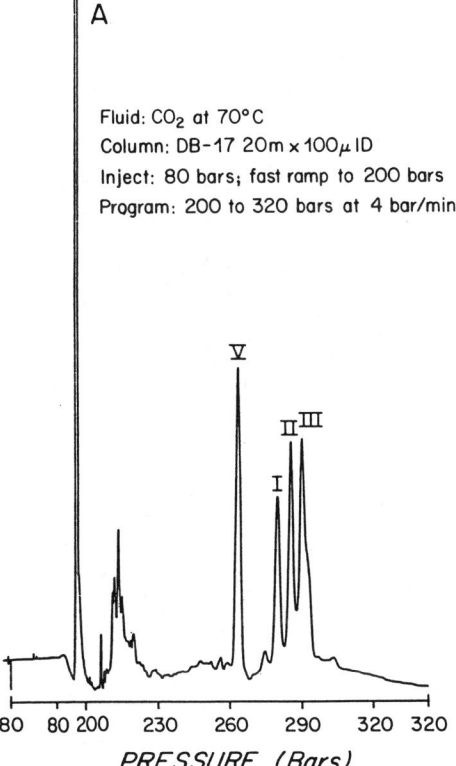

Figure 2A. SFC-FID chromatograms of carotenoids separated on DB-17 stationary phase. See Figure 1 for key to compound structures.

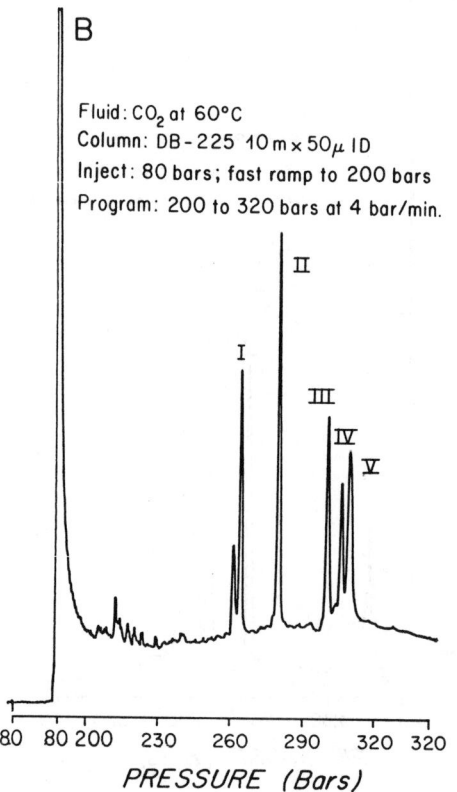

Figure 2B. SFC-FID chromatograms of carotenoids separated on DB-225 stationary phase. See Figure 1 for key to compound structures.

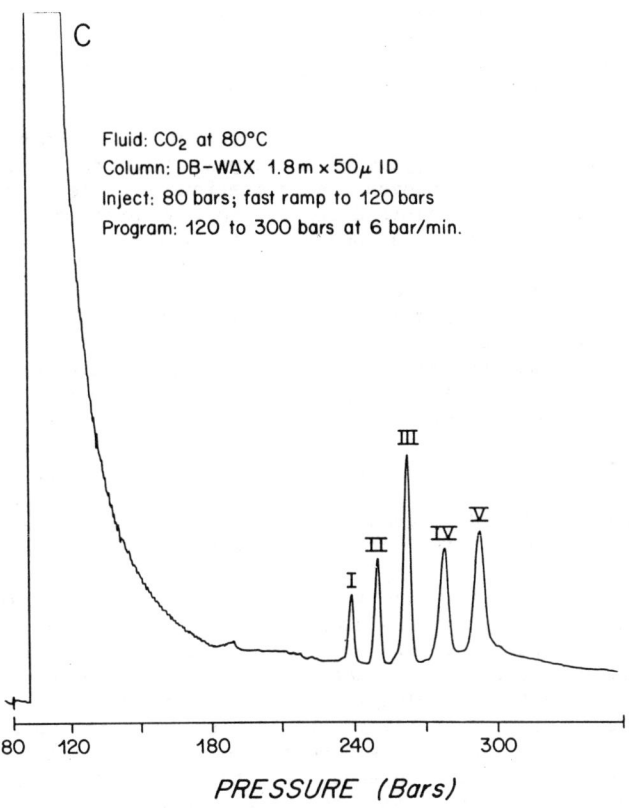

Figure 2C. SFC-FID chromatograms of carotenoids separated on DB-WAX stationary phase. See Figure 1 for key to compound structures.

selectivity is obtained with DB-WAX (a cross-linked Carbowax 20M). Figure 2C is a chromatogram obtained on a short (1.8 m x 50 μm ID) DB-WAX column using a very high linear flow velocity. These non-ideal conditions result in lower chromatographic efficiency, but are necessary for elution of the strongly retained polar carotenoids. The carotenoids I-V elute from DB-WAX with the expected order based on polarity.

We have also used longer (10 and 20 meter) DB-WAX columns using more optimal (lower) linear flow velocities, with attendent improvement in chromatographic efficiency. However, very high fluid densities (> 0.9 g/ml) and the use of polar modifiers (e.g. 2-5% MeOH) are required for elution of the more polar carotenoids from longer DB-WAX columns. Under these conditions, the stationary phase is not stable and continuous dissolution of the column coating leads to repeated plugging of the capillary restrictor. We conclude that further improvements in stationary phase cross-linking technology will be required for substantial progress in the use of polar modifiers for strongly polar solutes. The strong retention and high selectivity of DB-WAX make this phase the most promising one for the SFC separation of carotenoids and afford the possibility of using polar modifier gradients to enhance the quality of the chromatography.

It is important to emphasize that stationary phase selectivity is the dominant factor in the separation of these compounds, which differ primarily in end-group substitution. Other factors such as column diameter appear to be less important. The advantage of increased column efficiency obtained using smaller diameter columns (i.e. 50 μm) tends to be outweighed by the lower loading capacity, particularly for SFC-MS, because the signal/noise ratio is less favorable than for SFC-FID and higher amounts of analyte are required.

Due to their polarity and molecular weight, the successful elution of these compounds is crucially dependent on the design of the flow restrictor. We were not able to use small diameter (e.g. 5 micron) straight capillary restrictors because of pronounced solute precipitation and plugging problems. Instead, we fabricated jet-type restrictors directly on the columns using the method of Guthrie (34). The restrictors thus formed are mechanically sturdy and taper from the column internal diameter to approximately 1-3 microns over a length of 1 mm, providing nearly ideal decompression of the fluid. Significant advantages of the use of restrictors integral with the column are the ease of interfacing to the mass spectrometer, the absence of connecting dead volumes and active surfaces and the fact that the stationary phase extends virtually to the point of detection.

Mass Spectrometry. The use of supercritical CO_2 as a mobile phase provides an efficient means of transferring these labile compounds to the mass spectrometer ion source. Modifications to the Finnigan 4500 are minimal and fully compatible with normal GC-MS operation. Conversion between GC-MS and SFC-MS modes involves only an injector substitution and a column change. Figure 3 illustrates the reconstructed ion chromatogram obtained for the separation of

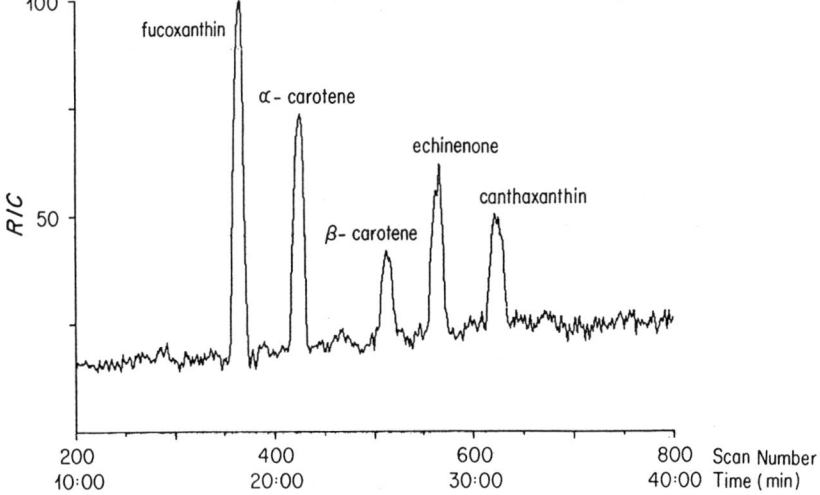

Figure 3. SFC-MS total ion chromatogram obtained for separation of a carotenoid test mixture on OV-17 (conditions as in Figure 2A).

a simple carotenoid test mixture on a 20 meter x 100 μm ID OV-17 column under conditions similar to Figure 2A. Adequate sensitivities (S/N = 5-10 for 50-100 ng injected on the column) are obtained using column effluent introduction normal to the quadrupole axis. However, we have not yet explored other ion source geometries with respect to optimizing sensitivity and better performance may be possible using axial effluent introduction. Ion source and analyzer pressure changes are substantial, increasing from 0.5 to 1.0 torr and from 3 to 6 x 10^{-5} torr, respectively, during pressure programming from 80 to 350 bars. However, these pressures are within the normal limits for chemical ionization. Background noise does not increase significantly. Observed CO_2 cluster ions (protonated) are not significant above the tetramer. We do not observe a significant effect on sensitivity due to increasing pressure, although Pinkston et al. (13) have reported lowered sensitivities for their SFC-MS system at analyzer pressures in the 10^{-4} torr range. In our experience, SFC pressure programming up to 400 bars using a properly dimensioned restrictor is compatible with normal mass spectrometer operating pressures. Several workers have reported that additional heating of SFC-MS restrictors is required as with FID detection (8, 36, 37). In the present work, mass chromatograms for the most polar carotenoids are observed to give smooth elution envelopes (e.g. fucoxanthin; cf. Figures 3 and 6), which suggests that particle formation during decompression is minimal without additional heating of the restrictor. It is possible, however, that we are detecting only that small percentage of the carotenoid analyte actually in the gas phase, the remainder forming particles to which the spectrometer is insensitive. Smith et al. (38) have studied restrictor performance in detail with respect to particle nucleation as a function of restrictor geometry, temperature, fluid flow rates and analyte volatility. However, effects of restrictor heating on thermolabile compounds have not been extensively evaluated. Pinkston et al. (13) have reported degradation of benzoyl peroxide used as a test compound when high restrictor temperatures were used for SFC-MS. Our attempts to use elevated restrictor temperatures with carotenoids resulted in reduced sensitivity, due to either analyte precipitation or to decomposition. Thus, although we generally operate our FID detector at 300-350°C, we are not convinced that this additional heating is required or beneficial in the case of MS detection, particularly when dealing with thermolabile components.

<u>Carotenes, Keto-carotenoids and Carotenoid Diols</u>. The CH_4-CI mass spectra of beta-carotene (I), echinenone (II), canthaxanthin (III) and astacene (IV) are shown in Figure 4. Under the soft ionization conditions used, these compounds exhibit extremely simple spectra consisting primarily of $[M+1]^+$. The usual $[M+29]^+$ and $[M+41]^+$ adduct ions are present but are of very low intensity. A small $[M+H+CO_2]^+$ peak is also observed for some of the compounds. The spectrum of the carotenoid diol, zeaxanthin (VIII, Figure 5), is also very simple, exhibiting a base peak, $[M+1]^+$, along with a prominent $[M+1-H_2O]^+$ ion. The presence

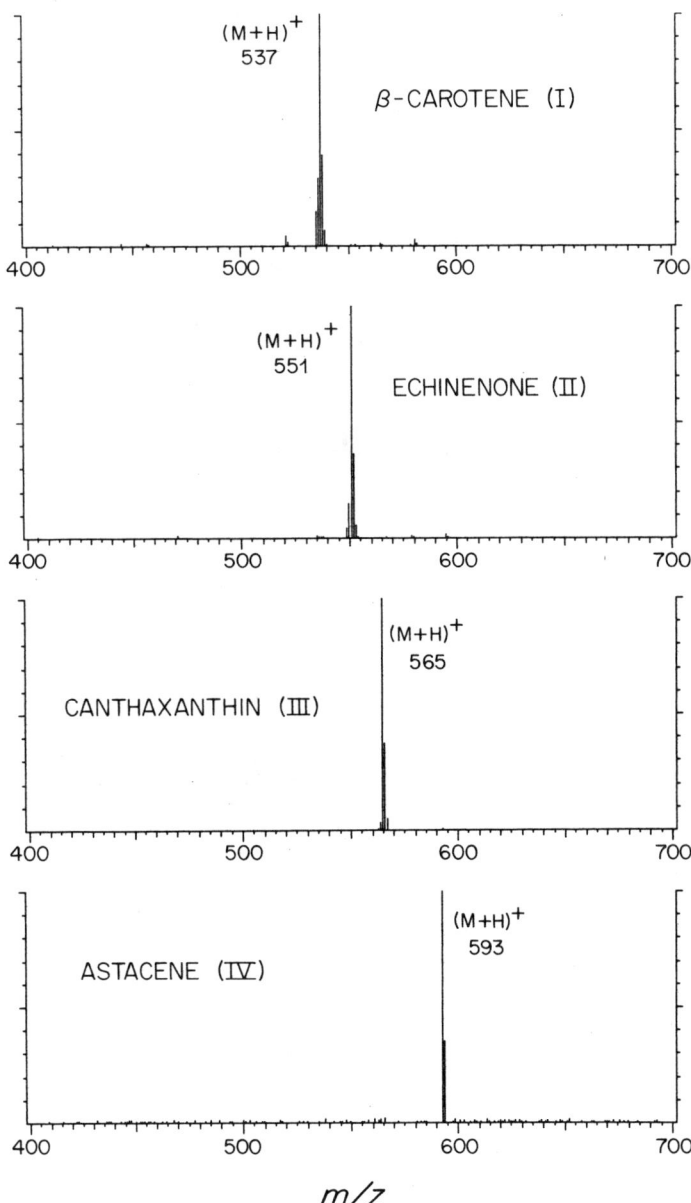

Figure 4. Methane chemical ionization mass spectra of several carotenoids. (Structures given in Figure 1.)

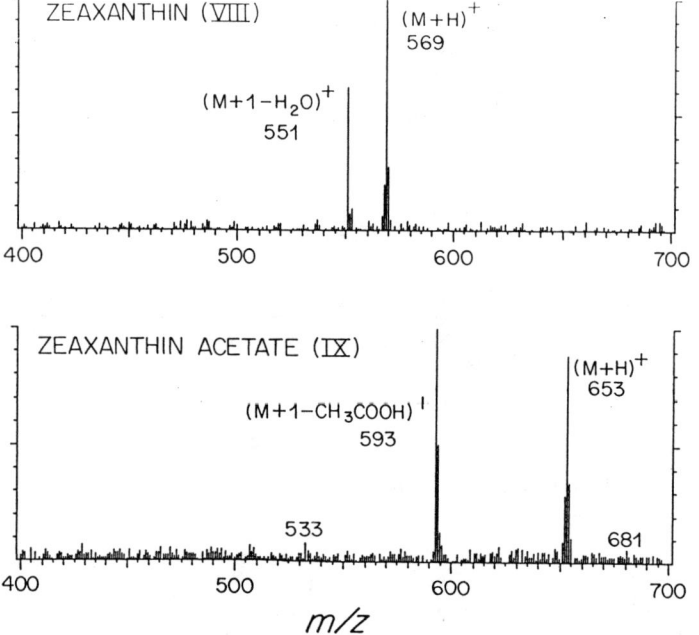

Figure 5. Methane chemical ionization mass spectra of the carotenoid diol, zeaxanthin and its acetylated derivative, zeaxanthin acetate.

of two hydroxyl groups is not apparent. The mass spectrum of the acetylated product, zeaxanthin diacetate (IX), however, exhibits an 84 dalton shift for the molecular ion and sequential losses of acetic acid, clearly indicating the structure to be a diol.

The simplicity of the carotenoid spectra under these conditions limits the structural information available to the analyst. As noted earlier, structural information regarding the carotenoid polyene chain and end-groups provided by electron ionization fragmentation, may arise largely from thermal degradation processes (31). Carnevale et al. (32), have demonstrated that similar structural information can be obtained using chemical ionization, although heated direct insertion probes were also used in their study. Further research is needed to determine the importance of thermal degradation processes (apparently minimized in SFC-MS) in providing structural information as opposed to processes initiated by ionization alone. There are several possibilities for creating more energetic ionization conditions, including higher source temperatures, the use of hydrogen (lower proton affinity) as a reagent gas or the use of mixed chemical ionization-charge exchange ionization. In recent work, we have noted increased fragmentation of the polyene backbone for some carotenoids (e.g. elimination of toluene) at high CO_2 pressures, apparently due to charge exchange. This suggests charge-exchange ionization as a means of producing significantly more fragmentation and we are actively exploring this area. It should be pointed out, however, that the lack of fragmentation can be used to advantage in the analysis of complex mixtures to obtain molecular ion abundances without the problem of interfering fragments. Minimal fragmentation would also be advantageous for tandem mass spectrometry of complex carotenoid mixtures, since simple molecular ions could be collisionally dissociated prior to the second analyzer to provide the necessary structural information.

Fucoxanthin and Related Pigments. For some types of pigments, notably the fucopigments, we observe considerable fragmentation even under these mild ionization conditions. Fucoxanthin (structure V), an abundant pigment in marine diatoms, is extremely fragile and susceptible to rapid degradation in the oceanic water column by various pathways including hydrolysis, dehydration and epoxide rearrangement. We have previously identified fucoxanthin and its degradation products using a combination of off-line HPLC and in-beam desorption chemical ionization (39). The quality of the spectra and detection limits obtained by this method vary considerably and are strongly dependent on factors such as probe heating rate, condition of the probe tip and distribution of sample on the tip. Here we compare results of the desorption CI probe method with some of our SFC-MS results. The SFC-MS analysis of a 100 ng sample of fucoxanthin is shown in Figure 6. Fucoxanthin elutes as a sharp chromatographic peak and produces a very high quality spectrum exhibiting an intense protonated molecular ion at m/z 659 (Figure 6). Other key diagnostic fragments (cf. Figure 7) include $[M+1-n18]^+$ at m/z 641 and 623, $[M+1-60]^+$ at m/z 599, $[M+1-60-n18]^+$ at m/z 581 and 563, $[M+1-170]^+$ at m/z 489, $[M+1-170-60]^+$ at m/z 429

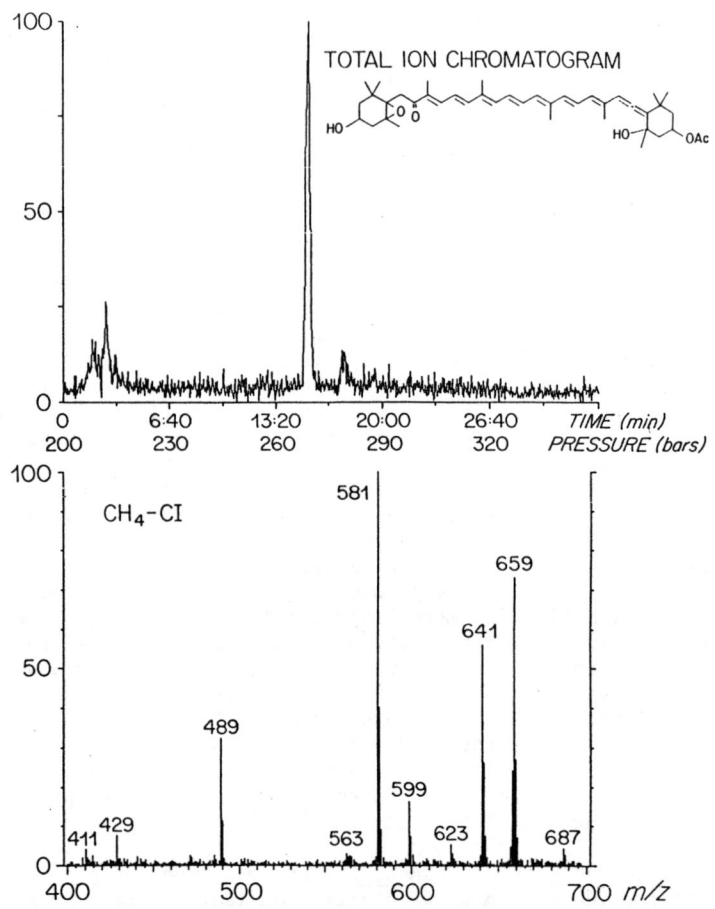

Figure 6. SFC-MS total ion current chromatogram and methane chemical ionization spectrum of fucoxanthin.

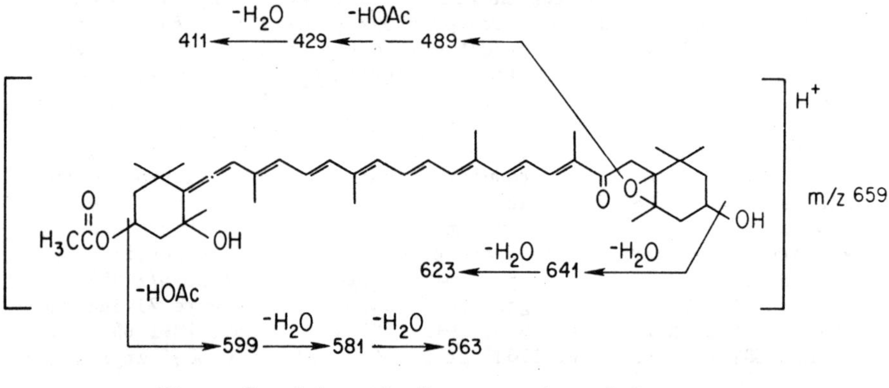

Figure 7. Schematic fragmentation of fucoxanthin.

and [M+1-170-60-18]+ at m/z 411. The effects of thermal degradation are absent and noise due to background is negligible. In contrast, Figure 8 shows results which were obtained by in-beam desorption CH_4-CI from a similar amount (100 ng) of fucoxanthin. The mass chromatograms for the [M+1]+ ion (m/z 659) and [M+1-18]+ ion (m/z 641) indicate the presence of both fucoxanthin and fucoxanthin dehydrate, raising the question of whether the dehydrate is present in the original sample or is produced by thermal dehydration on the probe. Furthermore, the mass spectra taken over different intervals exhibit significant variations and display extraneous fragments due to thermal degradation and reduced intensities for molecular ions. Thus, SFC-MS appears to be a more reliable means for interpreting fucopigment structures from relative intensity information.

Fragmentation patterns obtained by SFC-MS for the related fucopigments, isofucoxanthin (structure VI) and fucoxanthin-3-acetate (structure VII), are illustrated in Figure 9. The high quality of these spectra again reflect the gentle ionization conditions achievable with CO_2 SFC/CH_4-CI MS. As expected for isofucoxanthin, no ion is present at m/z 489, showing clearly that the 5,6-epoxy group is absent. Loss of ketene ([M+1-CH_2CO]+) is also apparent from the ion at m/z 617. Fragmentation of fucoxanthin-3-acetate produces relative intensities for the major ions (except for [M+1-$2H_2O$]+) which are very similar to those obtained for fucoxanthin.

Summary and Conclusions

The carotenoids, labile pigments which contain multiple functional groups spanning a range of polarities are normally considered to fall within the sphere of HPLC analysis. The preliminary results reported here show that separations of these compounds are also possible using supercritical fluid chromatography with CO_2 as the mobile phase and that this technique eventually may be competitive with HPLC. Using non-polar stationary phases, even relatively polar carotenoids can be eluted with pure supercritical carbon dioxide alone. However, the potential advantage of SFC in terms of column efficiency has not yet been attained experimentally in the case of the carotenoids. The practical resolution and selectivities observed for these compounds on the stationary phases studied are as yet rather limited compared with those observed for HPLC using bonded octadecyl or n-propyl amino columns, and are not yet adequate for dealing with complex carotenoid mixtures isolated from natural sources. Also, SFC-FID detection limits will need to be improved by a factor of 5-10 in order to match the low nanogram level currently possible with HPLC. Thus, from a chromatographic perspective, HPLC will remain the method of choice until improvements in SFC columns, as well as in SFC injection methods and detector design are achieved. Present limitations on the use of polar fluids and thus the polarity limits amenable to SFC-MS, imposed by the currently available selection and stability of polar stationary phases, will undoubtedly be eased by further research in state-of-the-art column technology. The

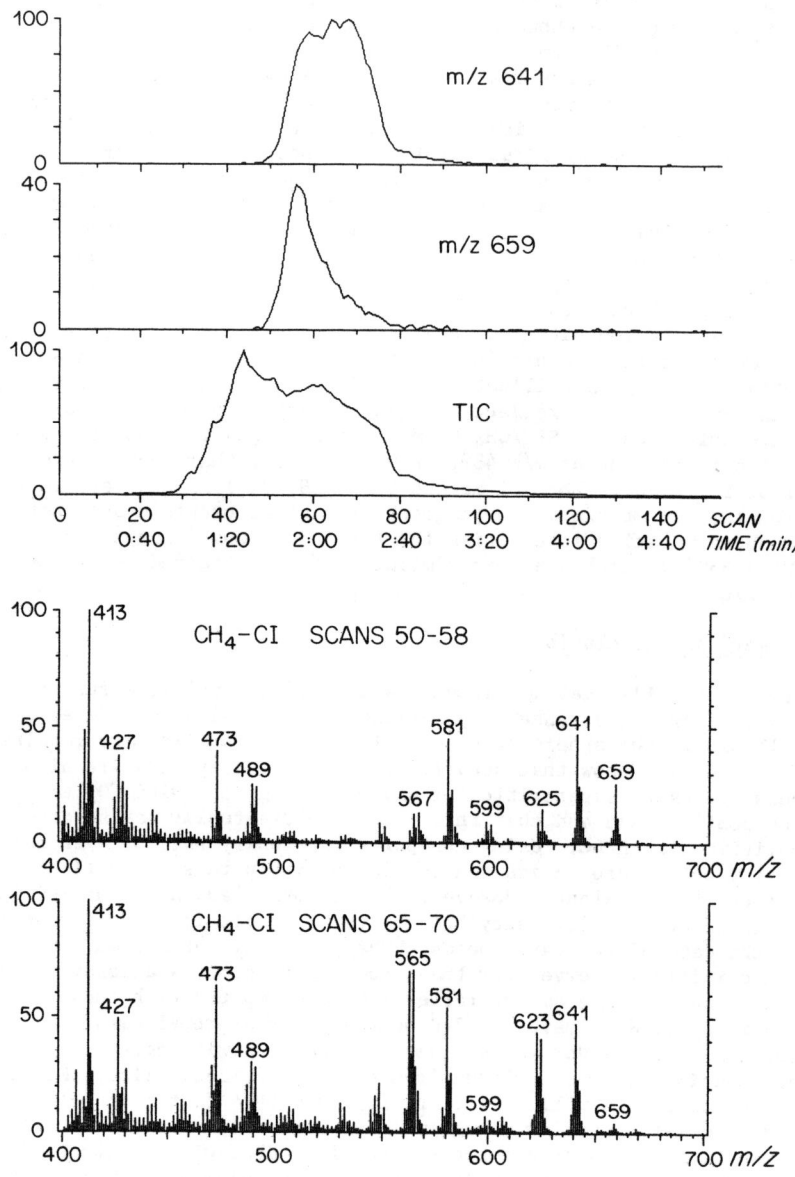

Figure 8. Total ion chromatogram, mass chromatograms and selected scans obtained for in-beam desorption CH_4-CI of fucoxanthin.

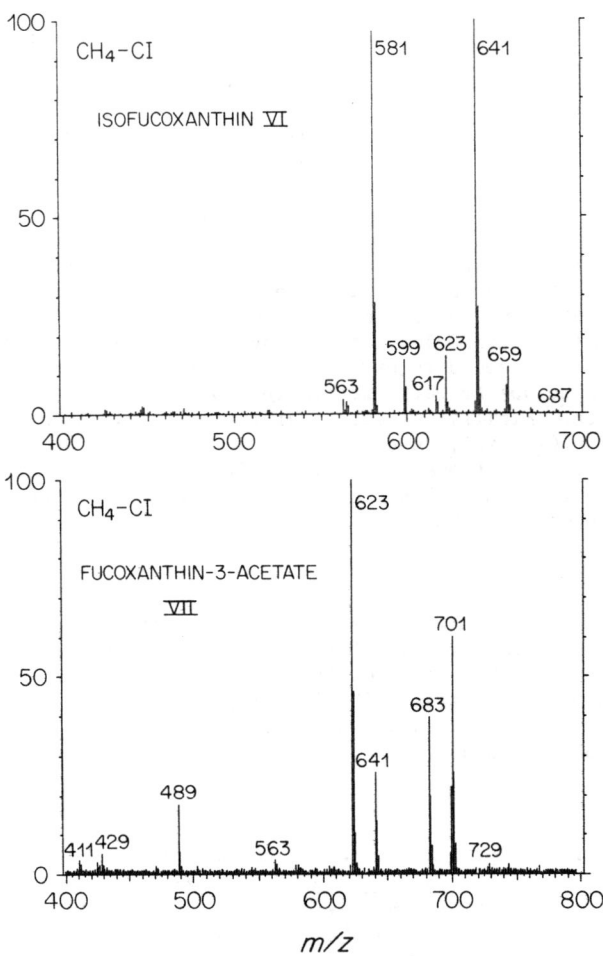

Figure 9. Methane chemical ionization mass spectra of the fucopigments isofucoxanthin and fucoxanthin-3-acetate.

most promising stationary phases for capillary SFC of complex natural mixtures of carotenoids appear to be cyanopropyl polysiloxane and polyethylene glycol phases.

A major advantage of SFC analysis of carotenoids is the compatibility of SFC with mass spectrometry. We have shown here that combined SFC-MS is a useful technique for the identification of these compounds in mixtures. In comparison with in-beam desorption chemical ionization techniques, the SFC-MS method produces superior quality spectra at similar sensitivity levels, particularly for very labile pigments such as fucoxanthin and its derivatives. The lower temperatures prevalent in the ion source due to the cooling effect of CO_2 expansion from the restrictor appear to result in extremely mild ionization conditions. With the exception of the fucopigments, CH_4-CI fragmentation of the other carotenoids studied is minimal. The simplicity of these spectra may be advantageous in determining low level distributions of these pigments using their molecular ion abundances. Further structural characterization would require more energetic ionization conditions or tandem mass spectrometry in combination with collision-induced dissociation. Finally, direct supercritical fluid injection-mass spectrometry (40, 41), involving no chromatographic separation, and therefore, short analysis times, offers a very attractive alternative for the analysis of pure carotenoid fractions and simple carotenoid mixtures.

Acknowledgments

We thank Drs. D. H. Repeta and J. Ertel for providing authentic carotenoid standards and for helpful discussions during the course of this work. This research has been supported by the Oceanic Chemistry Program, Office of Naval Research, under contract N00014-85-C-001. Woods Hole Oceanographic Institution Contribution No. 6510.

Literature Cited

1. Novotny, M.; Springston, S. R.; Peaden, P. A.; Fjeldsted, J. C.; Lee, M. L. Anal. Chem. 1981, 53, 407A-414A.
2. Peaden, P. A.; Fjeldsted, J. C.; Lee, M. L.; Springston, S. R.; Novotny, M. Anal. Chem. 1982, 54, 1090-1093.
3. Fjeldsted, J. C.; Lee, M. L. Anal. Chem. 1984, 56, 619A-627A.
4. Smith, R. D.; Felix, W. D.; Fjeldsted, J. C.; Lee M. L. Anal. Chem. 1982, 54, 1883.
5. Smith, R. D.; Udseth, H. R.; Kalinoski, H. T. Anal. Chem. 1984, 56, 2971.
6. Wright, B. W.; Udseth, H. R.; Smith, R. D.; Hazlett, R. N. J. Chromatogr. 1984, 314, 253.
7. Wright, B. W.; Kalinoski, H. T.; Udseth, H. R.; Smith, R. D. J. HRC&CC 1986, 9, 145-153.
8. Lee, E. D.; Henion, J. D. Paper 814, Pittsburgh Conference on Analytical Chemistry and Applied Spectroscopy, March 9-13, 1987, Atlantic City, NJ, USA.
9. Lee, E. D.; Henion, J. D. J. HRC&CC 1986, 9, 172-174.

10. Holzer, G.; DeLuca, S.; Voorhees, K. J. J. HRC&CC 1985, 9, 528-531.
11. Zaugg, S. D.; DeLuca, S. J.; Holzer, G. U.; Voorhees, K. J. J. HRC&CC 1987, 10, 100-101.
12. Henion, J. D.; Lee, E. D. Paper 671, Pittsburgh Conference on Analytical Chemistry and Applied Spectroscopy, March 9-13, 1987, Atlantic City, NJ, USA.
13. Pinkston, J. D.; Owens, G. D.; Burkes, L. J.; Chester, T. L.; Innis, D. P.; Delaney, T. E. Paper AGFD18, 193rd ACS National Meeting, April 5-10, 1987, Denver, CO, USA.
14. Chester, T. L. J. Chromatogr. 1984, 299, 424-431.
15. White, C. M.; Wagner, C. P.; Ravey, R. M.; Houck, R. K. Paper ANYL 48, 190th ACS National Meeting, September 8-13, 1985, Chicago, IL, USA.
16. Fjeldsted, J. C.; Jackson, W. P.; Peaden, P. A.; Lee, M. L. J. Chromatogr. Sci. 1983, 21, 222.
17. White, C. M.; Houck, R. K. J. HRC&CC 1985, 8, 293-296.
18. Jackson, W. P.; Later, D. W. J. HRC&CC 1986, 9, 175-177.
19. Schwartz, H. E.; Brownlee, R. G. J. Chromatogr. 1986, 353, 77-93.
20. Gere, D. R. Science 1983, 222, 253-259.
21. DeLuca, S. J.; Voorhees, K. J.; Langworthy, T. A.; Holzer, G. J. HRC&CC 1986, 9, 182-185.
22. Richter, B. E. J. HRC&CC 1985, 8, 299.
23. Smith, R. D.; Udseth, H. R.; Wright, B. W. J. Chromatogr. Sci. 1985, 23, 192-199.
24. Wright, B. W.; Kalinoski, H. T.; Smith, R. D. Anal. Chem. 1985, 57, 2823-2829.
25. Yonker, C. R.; Smith, R. D. Anal. Chem. 1987, 59, 727-731.
26. Yonker, C. R.; Smith, R. D. J. Chromatogr. 1986, 361, 25-32.
27. Hirata, Y. J. Chromatogr. 1984, 315, 39-44.
28. Levy, J. M.; Ritchey, W. M. J. Chromatogr. Sci. 1986, 24, 242-248.
29. Mourier, P.; Sassiat, P.; Caude, M.; Rosset, R. J. Chromatogr. 1986, 353, 61-75.
30. Repeta, D. J.; Gagosian, R. B. Geochim. Cosmochim. Acta 1987, 51, 1001-1009.
31. Vetter, W.; Englert, G.; Rigassi, N.; Schwieter, U. In Carotenoids; Isler, O., Ed.; Birkhauser Verlag: Basel, 1971; Chapter 4.
32. Carnevale, J.; Cole, E. R.; Nelson, D.; Shannon, J. S. Biomed. Mass Spectrom. 1978, 5, 641-646.
33. Watts, C. D.; Maxwell, J. R.; Games, D. E.; Rossiter, M. Org. Mass Spectrom. 1975, 10, 1102.
34. Guthrie, E. J.; Schwartz, H. E. J. Chromatogr. Sci. 1986, 24, 236-241.
35. McLaren, L.; Myers, M. N.; Giddings, J. C. Science 1968, 159, 197-199.
36. Smith, R. D.; Udseth, H. R. Anal. Chem. 1987, 59, 13-22.
37. Burkes, L. J.; Owens, G. D.; Lacey, M. P.; Simms, J. R.; Keough, T.; Pinkston, J. D. Paper 1046, Pittsburgh Conference on Analytical Chemistry and Applied Spectroscopy, March 9-13, 1987, Atlantic City, NJ, USA.

38. Smith, R. D.; Fulton, J. L.; Petersen, R. C.; Kopriva, A. J.; Wright, B. W. Anal. Chem. 1986, 58, 2057-2064.
39. Repeta, D. J. Ph.D. Thesis Woods Hole Oceanographic Institution, Woods Hole, MA, 1982.
40. Smith, R. D.; Fjeldsted, J. C.; Lee, M. L. J. Chromatogr. 1982, 247, 231-243.
41. Smith, R. D.; Udseth, H. R. Anal. Chem. 1983, 55, 2266-2272.
42. Liaaen-Jensen, S. In Carotenoids; Isler, O., Ed. Birkhauser Verlag: Basel, 1971; Chapter 3.

RECEIVED November 13, 1987

Chapter 13

Supercritical Fluid Chromatography with Fourier Transform Infrared Detection

Richard C. Wieboldt[1] and James A. Smith[2]

[1]Spectroscopy Research Center, Nicolet Instrument Corporation, Madison, WI 53711
[2]Winton Hill Technical Center, The Procter & Gamble Company, Cincinnati, OH 45224

> This paper describes the design of a packed column and a capillary column SFC/FT-IR system using a flow-through FT-IR analysis cell and their application to citrus oil and pyrethrin analyses. FT-IR detection with density programmed SFC separations is enhanced using a modified Gram-Schmidt algorithm. Infrared absorption bands due to the density changes in the supercritical carbon dioxide mobile phase are removed by spectral subtraction. FT-IR spectra collected across an unresolved chromatographic peak are used to resolve and quantitate the components of a peak in the citrus oil extract. Pyrethrin II is separated from a pyrethrin extract without thermal degradation using mild SFC conditions. FT-IR spectra have sufficient signal-to-noise ratio to clearly distinguish the cinerin and pyrethrin components.

Infrared spectroscopy is probably the most widespread analytical spectroscopic technique for identification and characterization of organic compounds. Because of this identification capability, infrared spectroscopy is desirable as a detection technique for chromatographic separations. With the advent of Fourier transform infrared spectroscopy, the speed and sensitivity of infrared detection is greatly enhanced making such applications feasible. FT-IR detection has been widely accepted as a detector for gas chromatography (GC/FT-IR) (1) and has been applied with limited success to liquid chromatography (LC/FT-IR) (2), and more recently to supercritical fluid chromatography (SFC/FT-IR) (3). The recent review articles cited here provide excellent introduction and references to current state-of-the-art in these areas.

In addition to identification, FT-IR is equally useful in its ability to function as a chemically specific detector for chromatographic effluents. The presence of absorption bands in certain regions of the infrared spectrum are characteristic of particular functional groups. For example, the C=O bond stretch (carbonyl functional group) causes an absorption band between 1825-1645 cm^{-1} in the mid-infrared spectrum. The exact position, say 1760 cm^{-1} versus 1720 cm^{-1}, serves to further distinguish between a carboxylic acid and a ketone. There are many such characteristic regions in infrared spectra, and it is the presence or absence of absorption bands in these regions which makes infrared so useful for structural elucidation.

The FT-IR spectrometer can be used to monitor any or several of these characteristic regions in real time. In this mode the FT-IR is not only collecting full infrared spectra for later identification, but it is also acting as a functional group specific detector. This is somewhat analogous to single ion monitoring in mass spectrometry with the advantage that an infrared absorption band is often more characteristic of a class of compounds than is a single mass ion.

FT-IR is often compared with mass spectroscopy as a chromatographic detector. Each technique provides unique information which is often complementary. IR provides clear chemical information while MS provides clear molecular weight and structural information. Taken together, the information greatly simplifies the determination of unknowns and almost always provides sufficient information for positive identification.

FT-IR Detection Methods. There are two methods for monitoring the infrared absorption of a chromatographic effluent: flow-through cell (4,5) and effluent deposition (6). Flow-through cell detection requires that the background remain constant throughout the experiment. With GC/FT-IR this is not a problem because the carrier gas stream is an inert gas having no infrared absorbance in the mid-IR. This requirement is a severe restriction for liquid chromatography where the mobile phase solvent typically has significant infrared absorption of its own. As a result, the usefulness of flow-cell LC/FT-IR is very limited.

The other method for obtaining IR spectra of chromatographic effluents is to deposit the eluted material on a surface, drive off the mobile phase, and examine the residual sample by FT-IR. This approach circumvents the mobile phase absorption problem but is cumbersome to carry out in practice. Techniques such as these have been employed for LC/FT-IR analysis (7,8,9) and demonstrated with SFC/FT-IR (10). One of the objectives of this study is to demonstrate the feasibility of using flow-through cell detection with SFC/FT-IR.

The success or failure of flow cell detection depends on the ability to cope with the absorption due to the supercritical fluid mobile phase. Because carbon dioxide is used as the supercritical fluid in the majority of applications, it is important to demonstrate its viability with FT-IR detection.

The absorption spectrum of supercritical carbon dioxide in the mid-infrared is compatible with the detection of most chemical species. Figure 1 shows the absorption bands for supercritical carbon dioxide in the mid-IR. The regions from 3800-3500 cm^{-1} and 2500-2200 cm^{-1} are totally blocked by the carbon dioxide and no useful information about sample absorption is available. A pair of weak absorption bands at 1381 and 1277 cm^{-1} are also present in the spectrum. These are caused by a Fermi resonance interaction between the O-C-O symmetric stretch, v_1, at 1330 cm^{-1}. The v_1 fundamental is not IR active which is why there is not a strong absorption band in this region.

The Fermi resonance absorption bands, which obviously are IR active, increase in intensity with density of the supercritical CO_2 fluid (11). This absorption becomes the limiting factor in selecting the pathlength for the flow-through analysis cell. In order to use this region of the IR spectrum, there must be enough analytical radiation remaining after the supercritical fluid absorbance to obtain adequate sensitivity for a particular application.

The supercritical CO_2 absorption bands change in intensity as a function of density but the band shape does not change - at least not at the 8 cm^{-1} spectral resolution typically used for this application. As a result, it is a simple matter to subtract the supercritical carbon dioxide absorption spectrum from an FT-IR data file collected during an SFC/FT-IR experiment. The subtraction factor is adjusted to exactly compensate for the Fermi resonance absorption. The resulting spectrum will then contain only absorption bands due to other components, if any, entrained in the supercritical fluid. The regions from 3800-3500 cm^{-1} and from 2500-2200 cm^{-1} appear as gaps in the spectrum because the supercritical carbon dioxide absorbs all the available infrared radiation in these regions.

Experimental

Two different SFC/FT-IR systems were used for this work. System 1 was used for the packed column analysis of volatile citrus oil components. System 2 was used for the pesticide and pyrethrin extract analyses.

<u>System 1</u>. This system consists of an HP1082B liquid chromatograph (Hewlett-Packard, Palo Alto, CA) modified for supercritical CO_2 operation, and a 60SX FT-IR spectrometer (Nicolet Instrument Corp., Madison, WI). The HP1082B system uses a manual backpressure regulator to maintain pressure throughout the SFC system. As a result, it is limited to applications which do not require density or pressure programming. The high pressure UV flow-through cell supplied with the chromatograph was used for FT-IR detection after replacing the standard quartz windows with infrared transparent zinc selenide windows.

This cell was placed in the 60SX microbeam sample compartment which uses 6x beam condensing optics and a narrow band MCT detector. The system allowed one infrared spectrum to be collected per second with 8 interferograms averaged per spectrum. The injector (Rheodyne #7520, 0.5 ul sample loop) and column oven were mounted next to the FT-IR spectrometer to reduce post-column dead volume.

<u>Citrus Oil Sample</u>. A synthetic test mixture of 10 volatile citrus oil components was prepared. The highest molecular weight

compound was d-limonene (136.24 g/mole) and the lowest molecular weight compound present was water. Limonene was 85% of the sample weight and the 0.5 ul sample was injected neat onto a 5 micron particle size PRP-1 column (Hamilton, 4 mm i.d. x 15 cm length). Supercritical carbon dioxide was pumped at a rate of 1 ml/min at a constant system temperature of 50°C. The column head pressure was 1750 psi while the system backpressure was maintained at 1400 psi.

System 2. This system consists of a Model 501 supercritical fluid chromatograph (Lee Scientific, Salt Lake City, UT) and a Nicolet 20 SXC FT-IR spectrometer. The SFC/FT-IR interface is a prototype design containing a 600 um I.D. by 5mm pathlength SFC/FT-IR flow-through cell, narrow band MCT detector, and optics designed to match the collimated beam from the main bench to the flow cell and detector.

The SFC/FT-IR flow cell is connected in-line at the end of the capillary SFC separation column and before the end-of-column restrictor. The column and restrictor are located in the chromatograph oven compartment. The flow cell is housed in an interface module between the spectrometer and chromatograph. It is connected by two lengths of 100 um I.D. deactivated fused silica tubing using a set of zero dead volume fittings (Scientific Glass Engineering, Austin, TX). The transfer lines and cell body were not heated for these experiments.

The 20SX FT-IR spectrometer was set up to collect 8 cm^{-1} resolution spectra with 8 scans coadded per file. Data files were stored on magnetic disk with a time resolution between files of 2.27 seconds.

Pesticide Sample. The pesticide mixture consists of aldicarb (5.6 mg/ml), methomyl (5.3 mg/ml), captan (5.0 mg/ml), and phenmedipham (5.1 mg/ml) prepared in dichloromethane. Samples were provided courtesy of Lee Scientific. This sample was separated using an SB-Methyl-100 10 meter 100 micron I.D. column with a 0.5 micron film (Lee Scientific). The chromatograph was programmed from an initial density of 0.180 g/ml to 0.360 g/ml at 0.010 g/ml/min after a 6.0 minute initial hold. The program rate was then immediately increased to 0.040 g/ml/min to a final density of 0.600 g/ml and held for 10 minutes. The oven temperature was maintained at 100°C throughout the experiment. The sample was delivered to the column using a 200 nL injection loop and a 22:1 split ratio.

Pyrethrin sample. The pyrethrin sample consists of 54 mg of a commercially available 20% pyrethrin extract, provided courtesy of Adams Veterinary Laboratory, prepared in 1.0 ml methanol. This sample was separated using the same column described above. The chromatograph was programmed from an initial density of 0.180 g/ml to 0.700 g/ml at 0.020 g/ml/min after a 6.0 minute initial hold. The final density was held for 5.0 minutes. The oven temperature was maintained at 100°C throughout the experiment. The sample was delivered to the column using a 200 nL injection loop and a 25:1 split ratio.

Results and Discussion

Comparison of FID and FT-IR Detection. Figure 2a shows the flame ionization detector (FID) chromatogram from the capillary SFC separation of the pesticide mixture. This trace was obtained without the SFC/FT-IR flow cell by connecting the capillary separation column directly to the end-of-column restrictor mounted in the FID. This serves as a reference to show the chromatographic separation obtained before connection to the SFC/FT-IR interface.

Figure 2b is the Gram-Schmidt reconstructed chromatogram (GSR) generated from the stored FT-IR data collected during the separation. Gram-Schmidt reconstruction is a technique used in GC/FT-IR applications to produce a chromatogram by comparing infrared data collected during a GC/FT-IR run with background data collected at the beginning of the run (12). The Gram-Schmidt response is completely non-specific; that is, it produces a response whenever there is any change in the infrared signal which is different from the background scans. In this case, the GSR responds primarily to the changing density of the supercritical carbon dioxide. The two programmed density ramps and the initial and final hold periods are clearly displayed. It is easy to detect the solvent peak at 13.23 minutes but difficult to detect the presence of the four pesticide peaks. Clearly some other approach is required to enhance detection.

Figure 3 shows the GSR (a) with a modified Gram-Schmidt reconstructed chromatogram (b) after compensating for the changing CO_2 density. The modification consists of including infrared data collected at the higher density in the set of background conditions used for the Gram-Schmidt calculation (13). Because high density supercritical CO_2 is now defined as part of the background conditions, the Gram-Schmidt algorithm ignores the changing supercritical fluid density. The result is a chromatogram which enhances the other chromatographic peaks and clearly reveals the elution of the four pesticides. For lack of any better term, this chromatogram is referred to as the Gram-Schmidt Plus reconstruction (GSP).

The primary function of the GSP chromatogram is to determine which of the hundreds of SFC/FT-IR data files contain FT-IR spectra collected during elution of chromatographic peaks. The reconstruction in Figure 3b, for example, indicates that FT-IR spectra for the peak at 24.09 minutes are contained in files 653-665. With this information one can quickly retrieve the FT-IR spectrum of this component.

Figure 4b shows the %T IR spectrum for the peak at 24.09 minutes. Because of the density programming, this spectrum was collected at a higher CO_2 density than that of the background scans obtained at the beginning of the run. As a result, the dominant spectral features are those of the carbon dioxide. Figure 4a is the spectrum of supercritical CO_2 (same as Figure 1) plotted to visually compare the two spectra. It is important to realize that the Gram-Schmidt Plus algorithm removes the CO_2 interference from the reconstructed chromatograms only. It does not affect the actual FT-IR spectra in any way.

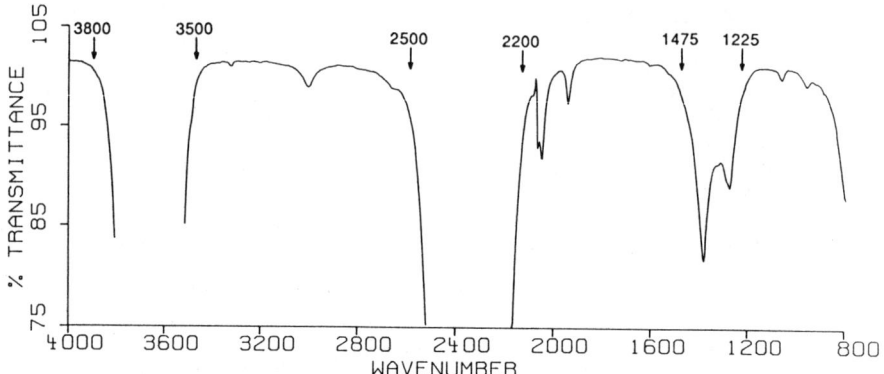

Figure 1. Infrared spectrum of supercritical carbon dioxide showing location of major absorption bands.

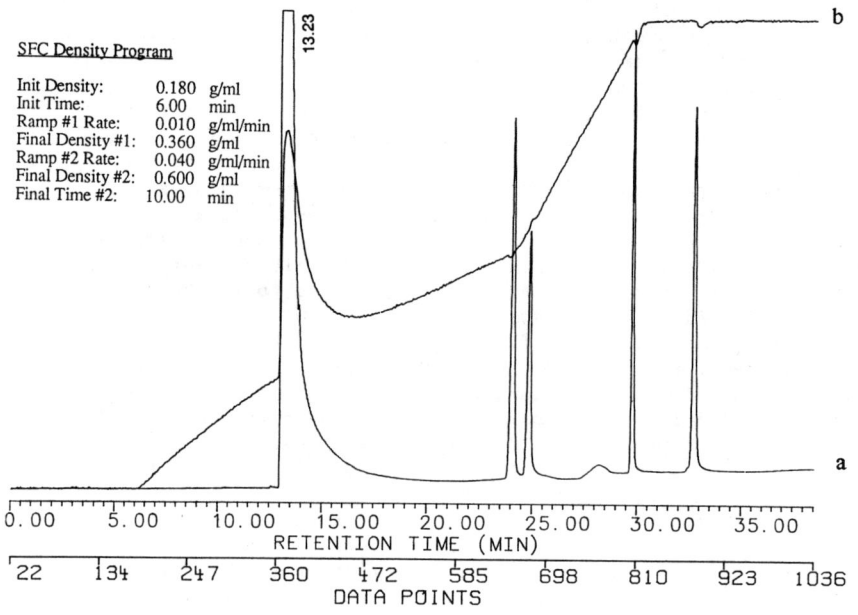

Figure 2. a) FID chromatogram from capillary SFC separation of pesticide mixture. b) Gram-Schmidt reconstructed chromatogram of the same separation.

13. WIEBOLDT AND SMITH *SFC with FTIR Detection*

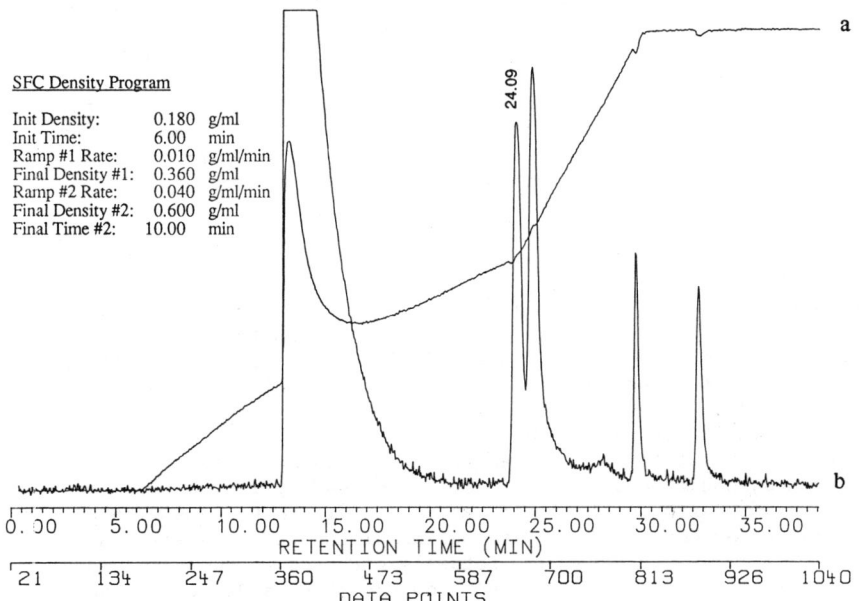

Figure 3. a) Gram-Schmidt reconstruction from Figure 3b. b) Gram-Schmidt Plus reconstructed chromatogram of the same separation.

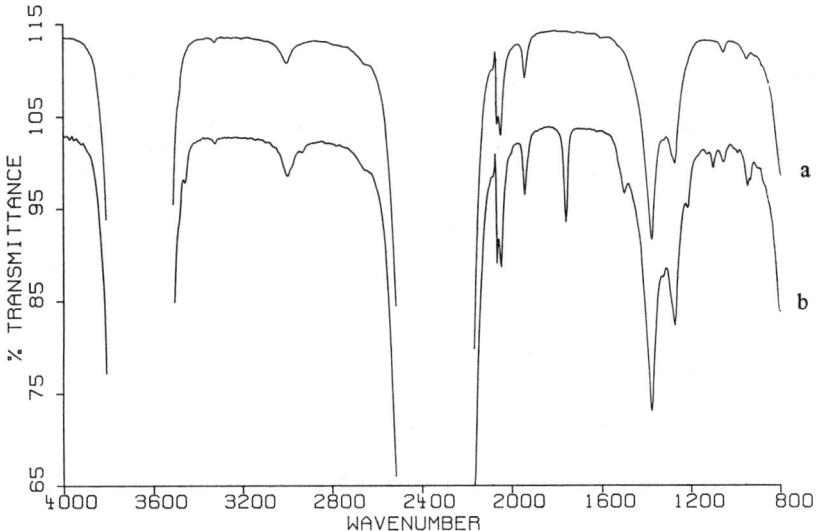

Figure 4. a) Infrared spectrum of supercritical carbon dioxide. b) FT-IR spectrum from peak at 24.09 minutes in pesticide SFC/FT-IR separation.

Because the spectral features of supercritical carbon dioxide are well defined, it is a simple matter to remove them, either manually or automatically, by spectral subtraction. Figure 5 is the spectrum from Figure 4b with the CO_2 absorbance removed. There is clearly sufficient spectral information to reveal the characteristic absorption bands which identify this component as aldicarb. It is important to point out that the upper blanked region of CO_2 does not block detection of the N-H stretch absorbance.

Analysis of Volatile Citrus Oil Components

Figure 6 is a Gram-Schmidt reconstructed chromatogram of a citrus oil components test mixture. The major peak in the chromatogram is due to limonene which is essentially a solvent. Note that the analysis time is extremely short, less than 5 minutes, and is accomplished under mild temperature conditions, 50°C. The resolution is not superb on all the compounds present. However, infrared detection can be used to resolve overlapping chromatographic bands. For example, the peak prior to limonene in the chromatogram (denoted with an *) appears to be pure with no shoulders evident. Close examination of the infrared spectra of the peak shows that the peak actually consists of two compounds. An infrared spectrum abstracted from the front side of the peak, File #452, matches the reference library spectrum of myrcene as is shown in Figure 7. While the infrared spectrum of File #464 from the backside of the peak, is a very good match with α-pinene (Figure 8). Thus, at least two compounds are present in the peak.

Pyrethrin Extract Analysis. Pyrethrins are a group of naturally occurring pesticides extracted from "Chrysanthemum cineraniaefolium." As with any pesticide, regulations require quantitative measurements and positive identification of the active ingredients. Currently, gas chromatography is used for the analysis. Pyrethrin I and II must be separated from the other components in the extract under mild temperature conditions. Pyrethrin II, which differs by only the presence of a terminal methyl ester group, rapidly decomposes at temperatures required to affect the separation. This makes quantitation exceedingly difficult. SFC is a logical alternative for this application because it uses much milder temperature conditions than GC. This eliminates the possibility of the sample degradation caused by temperature.

Figure 9 is the FID chromatogram of a capillary SFC separation of a commercially available pyrethrin extract. The active components are Pyrethrin I at 22.54 min and Pyrethrin II at 23.91 min. Both show good peak shape with no evidence of thermal decomposition. The group of peaks eluting before retention time 19 minutes are from the petroleum based extract matrix. The solvent peak at 8.79 min is methanol. The small peaks at 21.98 and 23.45 minutes are Cinerin I and II respectively. These are related components in the extract having the same structure as pyrethrin except for that the terminal methylene in the pyrethrins is replaced by a methyl group.

Figure 10 shows the SFC/FT-IR spectra for a) Cinerin II and

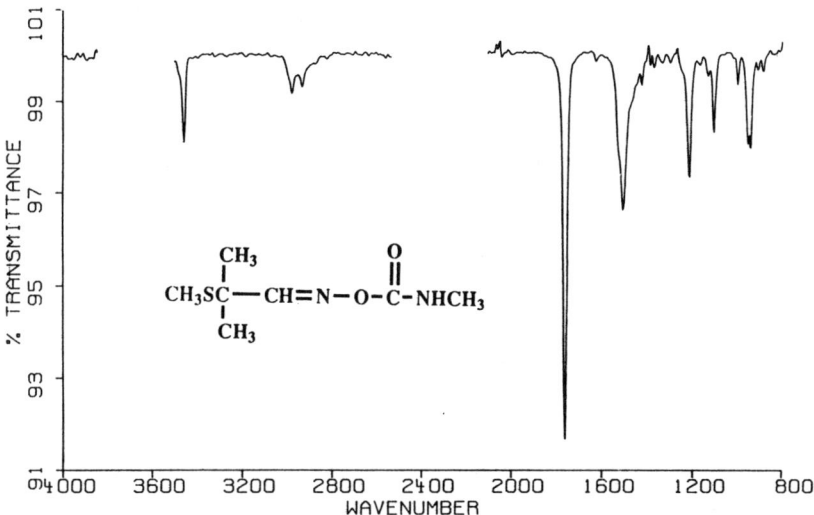

Figure 5. FT-IR spectrum from same peak as Figure 4b with supercritical carbon dioxide absorbance removed.

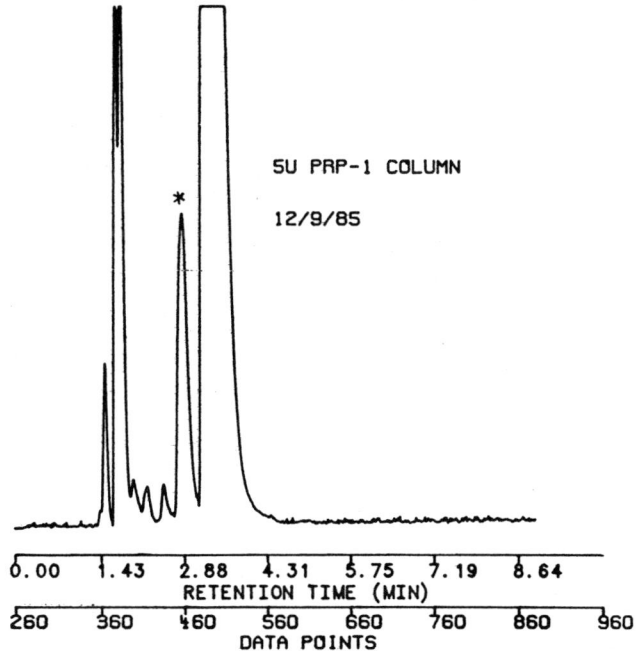

Figure 6. Gram-Schmidt reconstructed chromatogram of citrus oil test mixture.

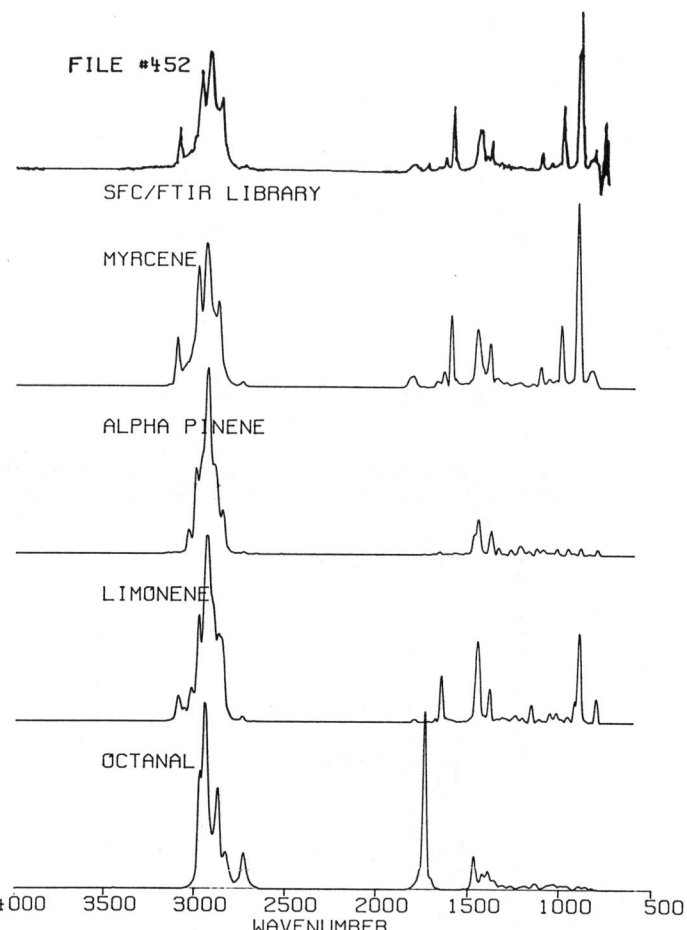

Figure 7. Spectral search library reference spectra results compared to File #452.

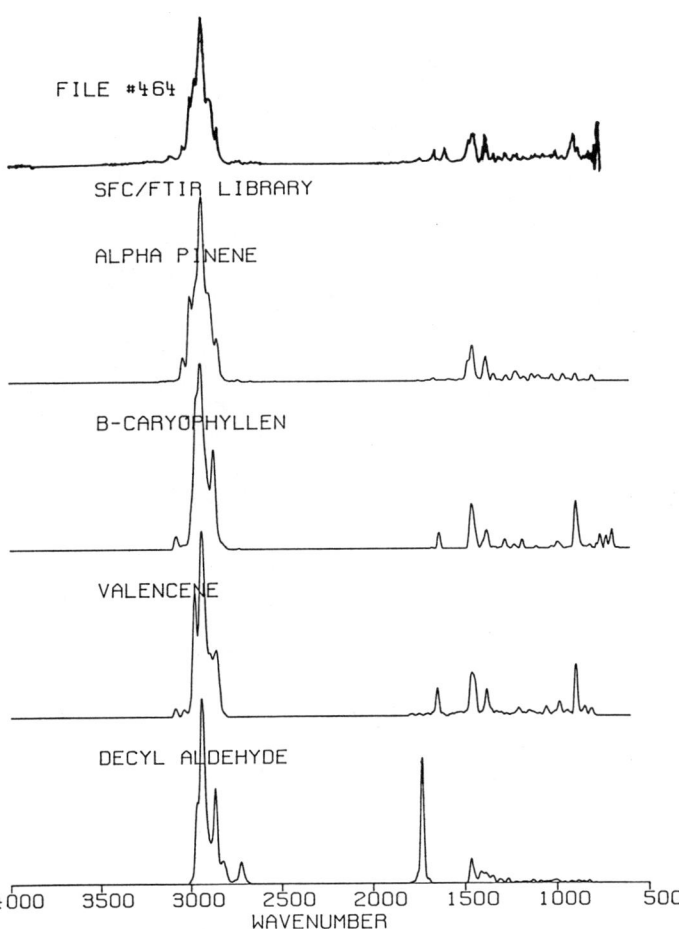

Figure 8. Spectral search library reference spectra results compared to File #464.

240 SUPERCRITICAL FLUID EXTRACTION AND CHROMATOGRAPHY

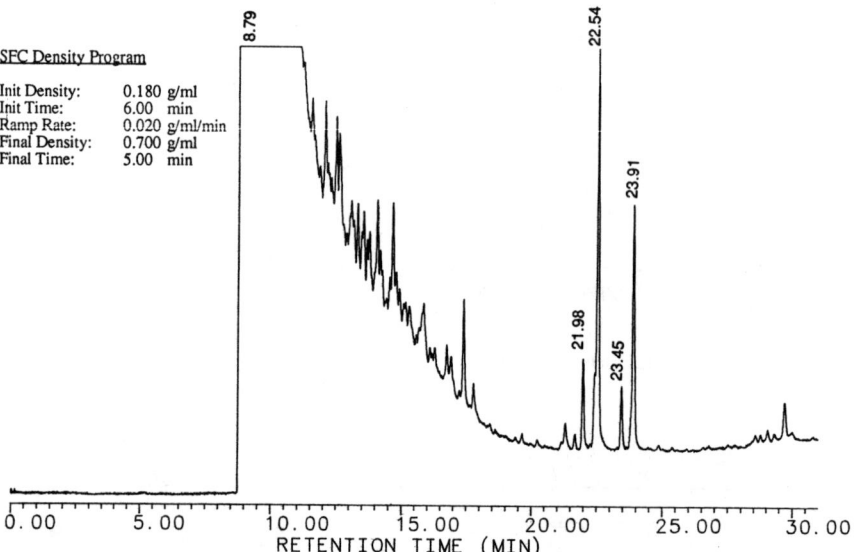

Figure 9. FID chromatogram from capillary SFC separation of pyrethrin extract.

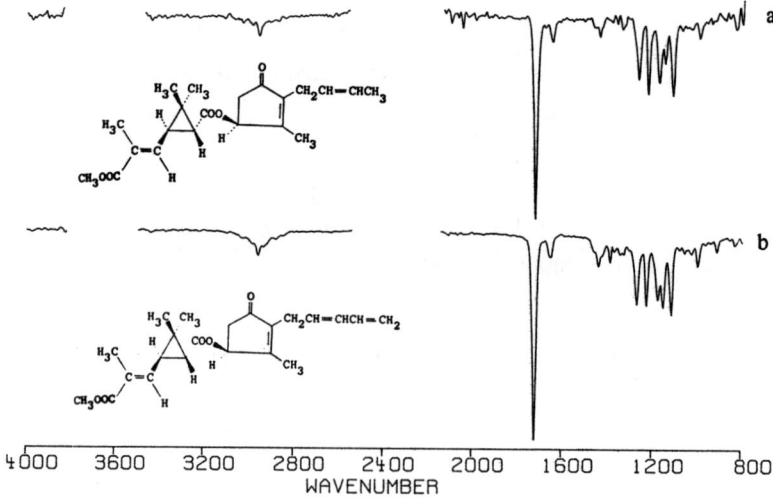

Figure 10. a) FT-IR spectrum for cinerin II peak at 23.45 minutes in pyrethrin SFC/FT-IR separation. b) FT-IR spectrum for pyrethrin II peak at 23.19 minutes.

b) Pyrethrin II and their structures. The II series are diesters having both a terminal methyl ester group and the ester linkage between the two rings. This gives rise to the complex C-O stretch absorbance bands between 1300-1100 cm^{-1}. The only feature which can be used to distinguish pyrethrin from cinerin is the C-H out-of-plan bending absorbance for the terminal methylene group in pyrethrin which occurs at 912 cm^{-1}. The presence of this absorption band is clear evidence of the terminal =CH_2 group.

Conclusions

These results demonstrate the feasibility of using a flow-through cell for FT-IR detection of carbon dioxide SFC effluent. The cell design is dictated by the restrictions imposed by the Fermi resonance absorbance of CO_2, not the chromatographic separation. As a result, the design works equally well for either packed-column or capillary SFC separations.

With packed column SFC separations, FT-IR detection can be useful in spectrally resolving components which may not be resolved by the chromatographic column. FT-IR also shows good sensitivity for the smaller quantities of materials encountered with capillary SFC applications. Spectral quality is sufficient for identification and for distinguishing subtle differences between related compounds.

Acknowledgments

The authors thank D. Wymer of the Procter & Gamble Company for assistance with the citrus oils SFC/FT-IR analyses, J. Freal of Adams Veterinary Research Laboratories for the pyrethrin extract, G. Adams and K. Kempfert of Nicolet Instrument Corporation for assistance with the SFC/FT-IR interface and pyrethrin extract analysis respectively.

Literature Cited

1. Herres, W. HRGC-FT-IR: Capillary Gas Chromatography - Fourier Transform Infrared Spectroscopy; Huthig: Heidelberg, 1987.
2. Hellgeth, J.W.; Taylor, L.T. J. Chromatogr. Sci. 1986, 24, 519-528.
3. Jinno, K. Chromatographia 1987, 23, 55-62.
4. Olesik, S.V.; French, S.B.; Novotny, M. Chromatographia 1984, 18, 489-495.
5. Johnson, C.C.; Jordan, J.W.; Taylor, L.T.; Vidrine, D.W. Chromatographia 1985, 20, 717-723.
6. Shafer, K.H.; Pentoney, S.L.; Griffiths, P.R. Anal. Chem. 1986, 58, 58-64.
7. Kuehl, D.T.; Griffiths, P.R. Anal. Chem. 1980, 52, 1394-1399.
8. Kalasinsky, K.S.; Smith, J.A.S.; Kalasinsky, V.F. Anal. Chem. 1985, 57, 1969-1974.
9. Fujimoto, C.; Jinno, K.; Hirata, Y. J. Chromatogr. 1983, 258, 81.

10. Pentoney, S.L.: Shafer, K.H.; Griffiths, P.R. J. Chromatogr. Sci. 1986, 24, 230-235.
11. Morin, P.; Caude, M.; Richard, H.; Rosset, R. Chromatographia 1986, 21, 523-530.
12. de Haseth, J.A.; Isenhour, T.L. Anal. Chem. 1977, 49, 1977-1981.
13. Wieboldt, R.C.; Hanna, D.A. Anal. Chem. 1987, 59, 1255-1259.

RECEIVED October 9, 1987

INDEXES

Author Index

Andersen, M. R., 179
Braddock, R. J., 109
Bromund, Richard H., 208
Burkes, L. J., 144,191
Campbell, E. R., 179
Chao, R. R., 89
Chen, C. S., 109
Chester, T. L., 144
Delaney, T. E., 144
Eissler, Robert L., 63
Frew, Nelson M., 208
Friedrich, John P., 63
Fulton, John L., 44
Innis, D. P., 144
Johnson, Carl G., 208
Keough, T., 191
King, Jerry W., 63
Knowles, D. E., 179
Kopriva, Andrew J., 44
Krukonis, Val J., 26

Lacey, M. P., 191
Later, D. W., 179
Liaw, Y. J., 89
Lira, C. T., 1
Marentis, R. T., 127
Nagy, S., 109
Nixon, L., 179
Owens, G. D., 144,191
Pinkston, J. D., 144,191
Porter, N. L., 179
Richter, B. E., 179
Rizvi, S. S. H., 89
Simms, J. R., 191
Smith, James A., 229
Smith, Richard D., 44,161
Temelli, F., 109
Wieboldt, Richard C., 229
Wright, Bob W., 44
Yonker, Clement R., 161

Affiliation Index

College of Wooster, 208
Cornell University, 89
Lee Scientific, Inc., 179
Michigan State University, 1
Nicolet Instrument Corporation, 229
Pacific Northwest Laboratory, 44,161
Phasex Corporation, 26

Pitt–Des Moines, Inc., 127
The Procter & Gamble Company, 144,191,229
U.S. Department of Agriculture, 63
University of Florida, 109
Woods Hole Oceanographic Institution, 208

Subject Index

A

Absorption spectrum, of supercritical carbon dioxide, 230–231,234f
Acetone, use as cosolvent, 14
Acid–base interactions, effect on solid solubility, 14
Acoustic streaming, definition, 50
Activated carbon
 determination of breakthrough volume, 83–84
 use as analytical sorbent material, 82
Adsorbates, breakthrough volumes, 70,71f,83t,84

Adsorbent(s)
  determination of breakthrough
    volume, 84,85t
  properties, 67t
Adsorption
  applications with SFE, 63–64
  cold-pressed oils, drawbacks, 113
  design of high-pressure system, 64
Alcohols in orange oil,
  concentration, 110–111
Aldehydes in citrus peel oil,
  concentration, 110

B

Benzoyl peroxide, analysis by SFC–MS, 198
Binary diffusion coefficients, in
  supercritical fluids, 21t
Binary supercritical fluid solution
  π^* vs. pressure, 172,174f,175
  retention behavior, 169–175
Binary systems, phase behavior, 11,12f
Botanical chemistry
  gas chromatogram of chemical
    components, 132,133f
  importance, 132
Botanical structure
  effect of flaking thickness on oil yield
    and recovery rate, 132,133f
  importance, 132
Breakthrough volume(s)
  calculation, 68–69
  determination
    on activated carbon, 83–84
    on resinous adsorbent, 84,85t
  for sorbates on XAD-2 resin, 74
  influencing factors, 70,73
  measurement, 64
  method of presentation, 82–83
  values for adsorbent resins, 83t,84
  vs. pressure, 69–73
  vs. time, 74

C

Capillary supercritical fluid
  chromatography
  analysis of celery seed oil, 181,185f
  analysis of cold-pressed grapefruit
    oil, 186,187–188f
  analysis of food components, 179
  analysis of liquid CO_2 extract of
    hops, 181,184f
  analysis of soybean oil
    sample, 180–181,182–183f
  applications, 179–180

Capillary supercritical fluid
  chromatography—*Continued*
  characteristics, 144
  experimental procedure, 180
Carbamate pesticides in parsley, analysis by
  capillary SFC, 186,189f
Carbon dioxide
  advantages in extraction, 92–93
  Clausius–Mossotti function, 5
  dielectric constant
    measurement, 5
    vs. pressure, 128,130f
  effect on breakthrough volumes, 73
  phase equilibria, theory, 127–131
  solubility
    vs. density, 128,129f
    vs. temperature, 128,130f
  solvent power, 3,4f
  ternary phase diagram, 28,29f
  See also Supercritical carbon
    dioxide
Carotenes, mass spectrum, 218,219f,221
Carotenoid diols, mass
  spectra, 218,220f,221
Carotenoids
  analytical methods, 211
  mass spectra, 218,219f,221
  SFC–FID
    chromatograms, 212,213–215f,216
  SFC–MS chromatograms, 216,217f,218
  structural information, 221
  structures, 209,210f
Celery seed oil, analysis by capillary
  SFC, 181,185f
Chemical process, scientific
  development, 2,4f
Chemical processing techniques,
  advantages, 2
Cholesterol
  SFC–MS response vs. pressure, 196,198
  structure, 196
Cholesterol extraction, application
  of SFE, 31
Chromatographic efficiency,
  of SFC–MS, 199
Chromatographic retention behavior,
  role of supercritical
    fluids, 78,81–82
Chromatographic void volume
  measurements, assessment
    of solvent-induced polymer
    swelling, 73
Chrysene extraction, rate in supercritical
  carbon dioxide, 52,53t
Citrus oil(s)
  analysis by SFC–FT–IR, 231–232
  applications, 109
  potential applications of SFE, 115,117
  synthesis, 109–110

INDEX

Citrus peel oils
  composition, 110
  folding processes, 111–113
Clausius–Mossotti function
  equation, 5
  for carbon dioxide, 5
Cod liver oil, fraction composition, 93,96t
Codfish oil esters, fatty acid
  profiles, 32,33t
Coffee decaffeination, application of
  SFE, 30–31
Cold-pressed grapefruit oil, analysis by
  capillary SFC, 186,187–188f
Cold-pressed oil(s)
  extraction, 111
  properties of components, 111,112t
Cold-pressed orange oil
  solubility isotherm, 115,116f
  vapor pressure vs. temperature of
    components, 113,114f,115
Commercial plant
  design
    mechanical design, construction, and
      start-up, 140,142–143
    process design and economic
      evaluations, 137–141
  factors influencing performance and
    economic efficiency, 131–132
  key processing conditions, 134–136
Complex mixture extraction, SFC
  comparison of hazardous waste
    samples, 49t,50,51f
Complex mixtures of natural products
  characterization methods, 208
  components, 236,237f
Continuous countercurrent extraction
  process, flow diagram, 34,36f,37
Corn sweeteners, silylated, characterization
  by SFC–MS, 202,204,205f
Critical phenomena, history, 1
Critical point, definition, 128
Critical pressure, definition, 128
Critical temperature, 128
Cybotactic region, definition, 172

D

Density
  dependence of interaction energy, 5
  dependence of refractive index, 6
  dependence of solute retention, 163
  dependence of solvent power, 5–6
Dielectric constant
  fluid, 5
  measurements for carbon dioxide, 5
  vs. pressure, 128,130f

Differential heats of adsorption
  calculation for sorbates, 81–82
  effect of supercritical fluids on
    retention behavior, 78,81–82
Diffusion coefficients, for supercritical
  fluids, 21t,22f
Disodium p-nitrophenyl phosphate
  hexahydrate, hydrolysis, 37–41
Drug-related problem, 199,202,203f
Dynamic mode of SFE system
  description, 119,121
  efficiency, 199

E

Effluent deposition, description, 230
Eicosapentanoic acid concentration
  by extraction profile, 34,36f
  by SFE, 31–37
Enhancement factor
  effect of solid-phase properties, 10
  linear behavior, 9–10,12f
  Poynting enhancement effect, 10–11,12f
Enthalpy of solute phase transfer
  definition, 166
  of binary fluid solvent, 169
  vs. density, 169,170f
Enzyme-catalyzed reactions in supercritical
  fluid
  flow reactor configuration, 37–38,39f
  gel permeation chromatogram, 38,40f
  reaction sequence, 38,39f
  transesterification reactions with
    lipases, 38
Enzymes, effects of supercritical fluids on
  stability, 38,41
Equilibrium reaction of urea, 98,99f,100
Equipment for supercritical fluid
  chromatography–mass spectrometry, 194
Ethyl caproate, breakthrough volumes vs.
  pressure, 70,72f
Extraction of cold-pressed oils,
  drawbacks, 111
Extraction of eicosapentanoic acid,
  profile, 34,36f
Extraction pressures, determination, 117
Extractions, number, 84

F

Factor of the reaction field, calculation, 3
Fatty acid profiles
  codfish oil esters, 32,33t
  menhaden oil esters, 32,33t

Fatty acids
  composition in refluxing system, 96,97t
  factors affecting stability, 100
Fish oil(s)
  applications, 89
  concentration by SFE, 31–37
  current methods of fractionation, 92
  factors influencing use in foods, 89–90
  omega-3 fatty acid composition, 90,91t
  SFE, 92
  source of polyunsaturated fatty acids, 90
Flame ionization detector (FID)
  chromatogram, SFC separation of
    pesticide, 236,237f
Flow-through cell detection,
  description, 230
Fluid dynamics, importance, 131
Fluid enhancement factor
  behavior, 10–11,12f
  definition, 10
Folding processes in citrus oil
  advantage, 111
  description, 111
  influencing factors, 111,112f
Food component analysis, by capillary
  SFC, 179–189
Food-related problems, 202,204,205f
Fourier transform infrared spectroscopy
  (FT–IR)
  applications, 229
  detection methods, 230–231,234f
  spectrum for pesticide, 236,237f
Fractionation, methods for fish oil, 92
Fractionation and standardization
  techniques
  multiple-stage fractional
    extraction, 140,141f
  two-stage fractional extraction, 137,140
Fragmentation
  of fucopigments, 223,225f
  of fucoxanthin, 221,222f,223
Free energy of solution of solute in the
  mobile phase, 145
Fucopigments
  fragmentation patterns, 223,225f
  structures, 210f, 223
Fucoxanthin
  in-beam desorption, 223,224f
  properties, 221
  schematic of fragmentation, 221,222f,223
  SFC–MS chromatogram, 221,222f
  structure, 210f,221
Fucoxanthin-3-acetate
  fragmentation patterns, 223,225f
  structure, 210f,223

G

Gas chromatography, advantages and
  disadvantages, 149

Gram–Schmidt Plus reconstruction
  chromatogram for CO_2, 233,235f
  description, 233
Gram–Schmidt reconstructed chromatogram
  description, 233,234f
  effect of CO_2 density, 233,234f

H

Hazardous waste samples
  gas chromatographic profiles, 50,51f
  SFE, 49t,50,51f
Heats of adsorption, differential—See
  Differential heats of adsorption
High-mass ions, detection by SFC–MS, 204
History of supercritical fluids, 26–41
Hops extraction, application of SFE, 30
Hot finger, definition, 32

I

In-beam desorption, of
  fucoxanthin, 223,224f
Infrared spectroscopy, advantages, 229
Interaction energy
  calculation, 6
  density dependence, 5,7f
  spherical nonpolarizable
    point dipole, calculation, 3,5
Interface probe, schematic, 192,193f
Interface tip temperature, effect on SFC–MS
  response, 196,197f,198
Ion source pressure, effect on SFC–MS
  response, 198–199,200f
Isofucoxanthin
  fragmentation patterns, 223,225f
  structure, 210f,223

J

Jet-type restrictors, construction, 216

K

Kamlet–Taft solvatochromic scale,
  description, 164
Ketones in orange oil, concentration, 111

L

Labile compounds, 198
d-Limonene, applications, 109
Lipases, transesterification reactions, 38
Liquid carbon dioxide, properties, 128

INDEX

Liquid carbon dioxide extract of hops,
  analysis by capillary SFC, 181,184f
Liquid solubility in supercritical fluids
  binary phase behavior, 16–19
  ternary phase behavior, 18,20f

M

Mass spectrometer tuning, effect on
  SFC–MS, 195–196
Mass spectrometry (MS)
  carotenes, 218,219f,221
  carotenoid diols, 218,220f,221
  carotenoids, 218,219f,221
  combination with capillary SFC, 191–205
  equipment, 194
Mechanical design construction
  and start-up of commercial plant
  cleaning between runs and
    products, 140,142
  construction schedule and
    start-up, 142–143
  maintenance and reliability, 140
  materials of construction, 142
  piping and valves, 142
  pretreatment and posttreatment, 142
  quick-opening closures, 142
  seals and gaskets, 142
Melting of solids
  phase behavior of solid–fluid
    system, 11,12f
  phase behavior of supercritical
    fluids, 11,13
Menhaden oil, fractionation with different
  mole ratios of urea, 100t,101,103f
Menhaden oil ethyl esters, fatty acid
  profiles, 34,35t
Methanol, use as cosolvent, 14
Methyl arachidate, preparation, 195
Microextraction cells, design, 55,57f
Mixture parameters, calculations, 163
Mobile-phase density
  effect on retention behavior, 146
  vs. height equivalent, 146,148f
Mobile-phase velocity
  vs. column efficiencies, 149,150f
  vs. height equivalent, 146,148f
Multicomponent systems, solid
  solubilities, 14,16
Myrcene, reference library IR
  spectrum, 236,238f
Myrosinase, effect of supercritical fluid on
  stability, 38,41

N

Naphthalene
  retention behavior, 166,167–168f
  solubility in supercritical fluid, 3,4f
Natural products, characterization by
  SFC–MS, 208–225

O

Off-line supercritical fluid extraction
  apparatus, 45,46f,48
  collection of effluents, 48
  complex mixture extraction, 49t,50
  high-pressure extraction vessel, 45,47f,48
  trace level extraction, 48t,49
Omega–3 fatty acids
  composition
    in refluxing system, 96,97t
    of fish oil, 90,91f
  yield
    from supercritical fatty
      acids, 97,98t
    with and without
      pretreatment, 101,103f
On-line supercritical fluid extraction–
  gas chromatography
  advantages, 53,55,61
  apparatus, 55,56–57f,58
  experimental procedures, 55,58
  qualitative extraction analyses, 58,59f
  quantitative extraction
    analyses, 58,60f,61t
Orange peel, qualitative extraction
  analyses, 58,59f
Organic solvent extraction, comparison to
  SFE, 134
Orifice size, 195

P

π^* Polarizability/polarity values
  vs. density, 172,173f
  vs. pressure, 172,174f,175
Partial molar volume of the solute,
  determination, 162
Pesticide, analysis by SFC–FT–IR, 232
Phase behavior
  solid–fluid systems, 11,12f
  ternary systems, 11,13f,14,15f
Pilot plant testing
  parameter verification, 135
  primary goal, 135
  schematic, 135,138f

α-Pinene, reference library IR
    spectrum, 236,239f
Polycyclic aromatic hydrocarbons, recovery
    from XAD-2 resin, 61t
Poly(dimethylsiloxane), characterization
    by SFC-MS, 202,205f
Polyethylene glycol derivatives,
    preparation, 195
Polyethylene glycol oligomers, SFC-MS
    separations, 198-199,200f
Polyunsaturated fatty acids, source, 90
Poynting correction, description, 10
Poynting enhancement factor
    behavior, 10-11,12f
    definition, 10
Pressure
    breakthrough volumes, 69-73
    discussion, 198-199,200f
    effect on solubility in supercritical
        fluids, 115,116f
    vs. temperature and physical
        state, 128,129f
Process development data
    analysis, 137,138-139f
    effect of temperature on two
        parameters, 137,138f
    extraction efficiency vs. time
        on-stream, 137,139f
Process development unit testing
    process parameters, 135
    schematic unit, 135,136f
Pulse chromatographic method,
    advantage, 66
Pyrethrin, analysis by SFC-FT-IR, 232
Pyrethrin extract analysis
    FID chromatogram of capillary SFC
        separation, 236,241f
    SFC-FT-IR spectrum, 236,240f

Q

Qualitative extraction analyses, of orange
    peel, 58,59f
Quantitative extraction analyses, of XAD-2
    resin, 58,60f,61t

R

Refluxing systems with variable temperature
    and pressure
    composition of fractions from cod liver
        oil, 94,96t
    fatty acid composition, 96,97t
    omega-3 free fatty acid
        concentration, 96,97t
    schematic, 93,95f,96,99f

Refractive index, density dependence, 6
Restrictor orifice size, effect on
    SFC-MS, 195
Retention behavior in gas chromatography,
    vs. mobile-phase density, 145-146,147f
Retention factor of solute
    vs. density, 166,168f
    vs. pressure, 166,167f
Retention mechanism in supercritical fluid
    chromatography
    effect of density, 169,172
    effect of polar modifier, 172,175
    effect of solvent strength, 175,176f
    polarizability/polarity
        values, 172,173-174f,175
Retention processes in supercritical fluid
    chromatography
    experimental procedure, 165
    retention factor vs. density, 166,168f
    thermodynamic theory, 162-164
Retention thermodynamics in supercritical
    fluid chromatography, theory, 162-164
Retention volumes, measurement, 68
Retrograde region, definition, 128

S

Scanning electron microscopy
    of Tenax resin, 74,76-77f,78
    of XAD-2 resin, 78,79-80f
    schematic, 192,193f
Screening unit testing
    primary goal, 134
    process parameters, 134-135
    screening unit, 134,136f
Selectivity of fractionation, effect of
    temperature, 93,95f
Sensitivity, of SFC-MS, 199,204,218
Silylated oligosaccharides, characterization
    by SFC-MS, 202,204,205f
Single-pass system for fractionation of
    fish oil, solubility of fish oil in
    supercritical carbon
    dioxide, 93,94f
Solid-fluid system, phase behavior, 11,12f
Solids in supercritical fluids,
    solubility, 3-9
Solids melting—See Melting of solids
Solid solubilities in multicomponent
    systems, solute-solute-solvent
    systems, 16
Solubility
    of liquids in supercritical fluids, 16
    of solids in multicomponent systems,
        solute-cosolvent systems, 14,16

INDEX

Solubility—*Continued*
  of solids in supercritical fluids, 3–9
  of solutes
    vs. solvent density, 128,129f
    vs. temperature, 128,130f
  problems in prediction, 6,9
Solubility behavior
  in solvents, 6,7f
  in supercritical fluids, 6,8f
Solubility enhancement—*See* Enhancement factor
Solute–cosolvent–solvent systems, solid solubilities, 14
Solute distribution coefficient
  vs. solute retention factor, 163
  vs. temperature, 163
Solute physical properties, effect on solubility in supercritical fluids, 113,114f,115
Solute retention
  vs. density, 163
  vs. pressure, 162
Solute–solute–solvent systems, solid solubilities, 16
Solvatochromic measurement of supercritical fluid solvation environments
  description, 164
  vs. susceptibility of solute, 164–165
Solvatochromic method, application, 175
Solvent extraction, mass transport steps, 131
Solvent power
  characterization by enhancement factor, 9–11,12f
  density dependence, 6
  solubility dependence, 6,7–8f
Sorbates
  breakthrough volumes, 73–74,75f,84,85t
  calculation of differential heats of adsorption, 81–82
  characteristics, 68
  role of supercritical fluids in retention behavior, 78,81–82
  van't Hoff plot, 78,81
Soybean oil, analysis by capillary SFC, 180–181,182–183f
Static mode of supercritical fluid extraction system
  complicating factors, 119
  schematic, 119,120f
Supercritical, definition, 1
Supercritical carbon dioxide
  absorption spectrum, 230–231,234f
  GSP chromatogram, 233,235f
  See also Carbon dioxide
Supercritical carbon dioxide extraction
  results of continuous-flow experiments, 121,123f,124

Supercritical carbon dioxide extraction—*Continued*
  system
    dynamic mode, 119,121
    schematic, 117,118f,119
    static mode, 119,120f
Supercritical fluid–adsorbate–adsorbent systems
  adsorbent properties, 67t
  calculation of breakthrough volume, 68–69
  experimental procedures, 66–69
Supercritical fluid chromatograph, description, 66
Supercritical fluid chromatography
  advantages, 149,208–209
  applications, 161,209
  chewing gum analysis, 155,157f
  elution of silylated oligosaccharides, 155,156f
  experimental methods, 211
  honeycomb extract analysis, 155,158f
  instrumentation, 149,151,211
  oligomer analysis, 151,152f
  poly(dimethylsiloxane) analysis, 151,153–154f
  prediction of retention, 165–175
  red and black pepper analysis, 155,159f
  retention thermodynamics, 162–164
  solvatochromic measurement of solvation environments, 164–165
Supercritical fluid chromatography–Fourier transform infrared (SFC–FT–IR) spectroscopy
  citrus oil sample, 231–232
  pesticide sample, 232
  pyrethrin sample, 232
  system
    for citrus oil analysis, 231
    for pesticide and pyrethrin extract analysis, 232
Supercritical fluid chromatography–mass spectrometry (SFC–MS)
  advantages, 209
  applications, 192
  chromatograms, 193f,194
  development, 191–192
  effect
    of interface tip, 195–196
    of ion source, 211–212
    of MS tuning, 212
    of restrictor, 212
  experimental methods, 233
  fragmentation, 233,234f
  instrumentation, 236,237f
  interface probe, 240f
  MS equipment, 194
  sample preparation, 194–195
  sensitivity, 199,218

Supercritical fluid extraction (SFE)
  advantages, 44–45,113
  applications, 30t,31,113
  combination with capillary gas
    chromatography, 53,55–61
  comparison to conventional organic solvent
    extraction, 134
  design protocol
    pilot plant testing, 135,138f
    process development unit testing, 135,136f
  enzyme-catalyzed reactions, 37–41
  fish oil concentration, 31–37
  flow diagram of continuous countercurrent
    extraction process, 34,36f,37
  misapplication, 27–28
  monotonic recovery, 28,30
  of botanical feedstocks, 132,133f,134
  of fish oil, refluxing systems, 93,95–97,99
  of urea preconcentrated samples, 98–103
  off-line analysis, 45–51
  potential citrus oil applications, 115,117
  promises of capabilities and
    applications, 27t
  system, 64,65f
  ultrasound analysis, 50,52–53
  use in fish oil fractionation, 90–107
  with a clathrate vessel
    effluent composition, 105–107
    schematic, 101,104f,105
  with temperature gradient column, 101,102f
  yields of omega-3 fatty acids, 97,98t
  See also Supercritical carbon dioxide
    extraction
Supercritical fluid mobile phases,
  characteristics, 191
Supercritical fluid processing, food
  applications, 127
Supercritical fluids
  characterization of solvent
    power, 9–11,12f
  effects on enzyme stability, 38,41
  factors affecting solubility, 113,114f,115
  history, 26–41
  liquid solubility, 16–20
  melting of solids, 11–15
  solubility prediction, 6,9
  use as mobile phases in
    chromatography, 179–180
Supercritical processes
  development, 2
  factors influencing application, 2–3
Susceptibility of solute vs. solvents
  solvatochromic behavior, 164–165
  temperature, 196,197f,198

T

Temperature
  effect on solubility in supercritical
    fluids, 115,116f

Temperature—*Continued*
  vs. pressure and physical state, 128,129f
  vs. solubility, 128,130f
Tenax resin, scanning electron microscopic
  photographs, 74,76–77f,78
Tenax–TA resin, breakthrough volumes vs.
  pressures, 69
Tergitol nonionic surfactant-9
  SFC–FID chromatogram, 199,201f,202
  structure, 199
Ternary systems, phase
  behavior, 11,13f,14,15f
Terpene fraction of citrus peel oils,
  properties, 110
Terpene hydrocarbons, extraction from
  cold-pressed citrus oils, 117
Thermally labile compounds, SFC–MS
  separation, 198
Thermodynamics of solute retention in SFC
  mechanism, 169
  theory, 162–164
  vs. temperature, 166
Trace level extraction, extraction
  comparison of activated carbon by
  Soxhlet and SFE, 48t,49
Transport properties of supercritical fluids
  diffusion coefficients, 21f,22f
  viscosity, 18,20f,21
Triton X-100
  SFC–MS response vs. tip
    temperature, 196,197f
  structure, 196

U

Ultrasonic supercritical fluid extraction
  advantages, 53
  apparatus, 50,52,54f
  of roasted coffee beans, 53,54f
  rates for various adsorbents, 52,53t
Unsaturated fatty acids, families, 90,91f
Urea, effect of pretreatment on yield of
  omega-3 fatty acids, 101,103f
Urea complexes, composition, 100t
Urea preconcentrated samples
  equilibrium reaction, 98,99f,100
  supercritical fluid extraction, 98–103

V

van't Hoff equation
  effect of mobile-phase solvation, 145
  GC retention vs. temperature, 145
van't Hoff plot, for sorbates, 78,81
Vapor–liquid behavior
  example
    of type I behavior, 18,19f
    of type III behavior, 16,18,19f
  types, 16–17f

INDEX

Vegetable oils, supercritical fluid
  extraction, 63–85

X

XAD–2 resin
  breakthrough volumes of
    sorbates, 73–74,75f

XAD–2 resin—*Continued*
  quantitative extraction
    analyses, 58,60f,61t
  scanning electron microscopic
    photographs, 78,79–80f
XAD–7 resin, breakthrough volumes for
  sorbates, 84,85t

Production by Cara Aldridge Young
Indexing by Deborah H. Steiner
Jacket design by Carla L. Clemens

Elements typeset by Hot Type Ltd., Washington, DC
Printed and bound by Maple Press, York, PA

Recent Books

Personal Computers for Scientists: A Byte at a Time
By Glenn I. Ouchi
276 pp; clothbound; ISBN 0-8412-1000-4

The ACS Style Guide: A Manual for Authors and Editors
Edited by Janet S. Dodd
264 pp; clothbound; ISBN 0-8412-0917-0

Silent Spring Revisited
Edited by Gino J. Marco, Robert M. Hollingworth, and William Durham
214 pp; clothbound; ISBN 0-8412-0980-4

Principles of Environmental Sampling
Edited by Lawrence H. Keith
458 pp; clothbound; ISBN 0-8412-1173-6

Writing the Laboratory Notebook
By Howard M. Kanare
146 pp; clothbound; ISBN 0-8412-0906-5

Chemical Demonstrations: A Sourcebook for Teachers
By Lee R. Summerlin and James L. Ealy, Jr.
192 pp; spiral bound; ISBN 0-8412-0923-5

Phosphorus Chemistry in Everyday Living, Second Edition
By Arthur D. F. Toy and Edward N. Walsh
362 pp; clothbound; ISBN 0-8412-1002-0

Chemical Reactions on Polymers
Edited by Judith L. Benham and James F. Kinstle
ACS Symposium Series 364; 483 pp; ISBN 0-8412-1448-4

Catalytic Activation of Carbon Dioxide
Edited by William M. Ayers
ACS Symposium Series 363; 212 pp; ISBN 0-8412-1447-6

Pharmacokinetics: Processes and Mathematics
By Peter G. Welling
ACS Monograph 185; 290 pp; ISBN 0-8412-0967-7

Polynuclear Aromatic Compounds
Edited by Lawrence B. Ebert
Advances in Chemistry Series 217; 396 pp; ISBN 0-8412-1014-4

For further information and a free catalog of ACS books, contact:
American Chemical Society
Distribution Office, Department 225
1155 16th Street, NW, Washington, DC 20036
Telephone 800-227-5558

Vicente Sanchez
Department Of Chemistry
VIRGINIA TECH
Blacksburg, VA 24061